Industrial Maintenance: Mechanical Fundamentals and Practices

Industrial Maintenance: Mechanical Fundamentals and Practices

Donald Wilkins

NY RESEARCH
P R E S S

New York

Published by NY Research Press
118-35 Queens Blvd., Suite 400,
Forest Hills, NY 11375, USA
www.nyresearchpress.com

Industrial Maintenance: Mechanical Fundamentals and Practices
Donald Wilkins

International Standard Book Number: 978-1-63238-694-6 (Hardback)

Cataloging-in-Publication Data

Industrial maintenance : mechanical fundamentals and practices / Donald Wilkins.
    p. cm.
Includes bibliographical references and index.
ISBN 978-1-63238-694-6
1. Plant maintenance. 2. Industrial equipment-- Maintenance and repair.
3. Machinery--Maintenance and repair. I. Wilkins, Donald.
TS192 .I53 2019
658.2--dc23

Contents

Preface VII

Chapter 1   **Introduction to Industrial Maintenance**..................................... 1

    i. Maintenance   1

    ii. Maintenance Engineering   23

    iii. Maintenance Management   24

    iv. Maintenance Testing   25

    v. Machine Element   25

Chapter 2   **Industrial Failures** 37

    i. Failure Analysis   38

    ii. Failure Cause   42

    iii. Failure Mode, Effects, and Criticality Analysis   63

    iv. Failure Rate   69

    v. Fault Reporting   70

Chapter 3   **Lubrication** 73

    i. Friction   73

    ii. Wear   86

    iii. Lubrication   91

    iv. Grease Fitting   97

Chapter 4   **Bearings** 100

    i. Ball Bearings   101

    ii. Plain Bearing   107

    iii. Roller Bearing   119

    iv. Air Bearing   122

    v. Spiral Groove Bearing   126

Chapter 5   **Rigging** 142

    i. Synthetic Rigging   142

    ii. Sailboat-Running Rigging   146

    iii. Standing Rigging   148

Chapter 6   **Fasteners** 150

    i. Mechanical Fasteners   157

    ii. Threaded Fasteners   158

Chapter 7   **Maintenance Practices Across Varied Industries** **164**

      i. Aircraft Maintenance   164

     ii. Automotive Maintenance   167

    iii. Bridge Maintenance   171

    iv. Firearm Maintenance   173

     v. Army Engineering Maintenance   179

Permissions

Index

Preface

The smooth functioning of an industrial plant requires a number of corrective, preventive and predictive maintenance strategies. Such strategies involve operational and functional checks, repair and replacement of components, servicing, building infrastructure, etc. Preventive maintenance is performed with the intent of avoiding failures, production costs, losses and safety violations. Corrective maintenance is applied to malfunctioning equipment, and may involve processes such as welding, metal flame spraying, etc. Due to the advent of sensing and computing technology, predictive maintenance has become a possibility. It involves sensors to monitor key parameters of the system health and predict any breakdown before it happens. This textbook is a compilation of chapters that discuss the most vital concepts in the field of industrial maintenance. Different approaches, evaluations and methodologies in this field have been included in this book. It will serve as a reference to all professionals and students associated with this field.

A foreword of all chapters of the book is provided below:

Chapter 1- Industrial maintenance includes a variety of activities and processes such as servicing, functional checks, repair or replacement and building infrastructure, besides others. The aim of this chapter is to provide an introduction to industrial maintenance, through an elucidation of the topics like maintenance, maintenance engineering, maintenance management and maintenance testing;

Chapter 2- It is vital to ensure the appropriate functioning and operation of machinery and manufacturing processes for continued production. All the vital aspects of industrial failures that may occur in the operation of industrial plants have been carefully analyzed in this chapter, such as failure analysis, failure cause, fault reporting, failure rate, failure mode, effects, and criticality analysis;

Chapter 3- A lubricant is used to reduce friction and wear between two surfaces in contact. It can be a solid, liquid or gas. Adequate lubrication is essential for the smooth and continuous operation of machines and for prevention of excess stresses at bearings. This chapter has been carefully written to provide a comprehensive understanding of the stresses and strains that the different parts of a machine are subject to owing to friction and the lubrication that can be applied to reduce such effects. It analyzes the fundamental concepts of grease fitting, friction and wear, etc.;

Chapter 4- A bearing is an element of a machine that either allows or prohibits motion of machine elements or reduces friction between moving parts. The topics elaborated in this chapter address the different types of bearings that are used in machinery, such as plain bearing, ball bearing, roller bearing, air bearing and spiral groove bearing;

Chapter 5- Rigging refers to the cables, ropes and chains that support a sailing ship or the masts of a sailboat. It is grouped into two categories, standing rigging and running rigging. This chapter discusses in extensive detail about the different types of rigging, such as synthetic, sailboat-running and standing rigging;

Chapter 6- A fastener is a hardware device, which affixes or joins objects together. Fasteners create non-permanent joints. Some of the varied fasteners used in machinery are mechanical fasteners, threaded fasteners and screws, which have been discussed in depth in this chapter;

Chapter 7- Maintenance typically involves adopting a set of preventive and scheduled practices for the efficient operation of equipment. This chapter is structured in such a way that it will provide an understanding of the maintenance practices required across varied industries, such as aircraft, automotive, bridge, firearm and arm engineering industries.

At the end, I would like to thank all the people associated with this book devoting their precious time and providing their valuable contributions to this book. I would also like to express my gratitude to my fellow colleagues who encouraged me throughout the process.

Donald Wilkins

Introduction to Industrial Maintenance

Industrial maintenance includes a variety of activities and processes such as servicing, functional checks, repair or replacement and building infrastructure, besides others. The aim of this chapter is to provide an introduction to industrial maintenance, through an elucidation of the topics like maintenance, maintenance engineering, maintenance management and maintenance testing.

Maintenance

Maintenance is defined as the compilation of tasks carry out in an installation, in order to avoid, analyze and outweigh the degradation of time or usage that they provoke in the equipment and systems which form that installation.

Maintenance department of an installation has six objectives which must mark and run its work:

- Assure the security of the installation for people, meaning that the normal or abnormal functioning does not endanger people. Of course, this objective is not sole for maintenance, but for the whole organization.

- Assure that the installation is secure for environment, so that a situation where the environment is damaged cannot happen. Of course, this objective is also not sole for maintenance, but for the whole organization.

- Achieve a certain value of availability.

- Achieve a certain value of reliability.

- Guarantee a long shelf life, at least according with the plant's repayment term.

- Achieve all of that, trimming the budget given. Often, the optimal budget for the maintenance of that installation.

Secureness Objective

Many of the accidents that occur in the installations happen because these suffer from functioning failures that turns them insecure. The vast majority of the installations are do managed according to manufacturer's instructions and the engineering that designed them, and safety standards established with the help of risk analysis are fulfill. These are installations with a probability of accident almost negligible. Only a failure in a component, equipment or system, changes the controlled conditions and makes the probability of an accident unacceptable. Thus, the first maintenance objective must assure that degradation that usage and time make on equipment outweigh enough to not risk people's security.

Environmental Objective

Likewise, many of the environmental accidents and incidents that cause a negative impact in environment happen due to an unexpected functioning of these, occurring from a failure in a component, equipment or system. The majority of the installations are do managed according to manufacturer's instructions and the engineering that designed them, and standards established with the help of environment risk analysis are fulfill. These are installations with a probability of environmental accident or incident almost negligible. In the majority of the installations, only a failure in a component, equipment or system, changes the controlled conditions and makes the probability of an environmental accident unacceptable. Therefore, the second objective of maintenance must assure that degradation that usage and time make on an equipment outweigh enough without risk of the environment suffering a negative impact.

Availability Objective

The availability of an installation is defined as the proportion of time when that installation has been in willingness of production or been used, regardless of whether it has or has not been used, for reasons beyond its technical condition.

The most important maintenance objective is to guarantee that the installation is in disposition of produce or be used a certain minimum number of hours per year. It is an error to think that the objective of maintenance is to achieve the maximum availability possible (100%) as this can become very expensive. Thus, achieving the marked availability objective with a certain cost is generally enough.

As will be seen further on, availability is an indicator that offers great possibilities of reckoning. Therefore, only for certain installations that operate intermittently (plants which produce in campaign, plants or installations that only function in certain periods) the objective of availability focuses on the hours reckon to produce, and the availability or not of the plant when it is not require to produce is relatively trivial. The definition of availability calculation estimation plays a crucial role in judging if the department of maintenance of any industrial installation is doing its job accurately or an improvement is needed.

The principal factors to bear in mind in the estimation of availability are the following:

- Total number of production hours.

- Total number of unavailability hours that might be caused due to different types of performance of maintenance.

- Programmed maintenance interventions that require plant stoppage.

- Programmed corrective maintenance interventions that require plant stoppage or load reduction.

- Non-programmed corrective maintenance interventions that unexpectedly stop production and thus, they had an incident in the planning already made in energy production.

- Partial unavailability number of hours, meaning, the number of hours when the plant is able to produce but with availability lower than nominal due to the poor condition of one section of the installation, that hinder it to work under full load.

Regarding to the acceptable values of availability in several types of industrial installations, achieving the availability objectives higher than 92% in a sustainable way (one or various years may be achieved, but not a continuous way) is quite an ambitious objective. Usually, industrial installations seek out for their goals between 99% (for the most demanding installations) and 50% in the less demanding cases when providing a much higher production than the market is able to assimilate.

It is crucial to highlight that the IEEE develop an specific rule for the reckoning of the installation that can be extrapolated to other type of installations, trying to avoid partial interpretations that may benefit some part at the expense of another (contractor owner, and so).

Reliability Objective

Reliability is an indicator that measures the capacity of a plant to achieve the production plan intended. In an industrial installation, it usually refers to fulfillment of the production intended, and, in general, understood by both internal and external clients. The achieving of this load program may have consequences like financial penalties, and hence the importance of measuring these value, and bearing it in mind when designing the maintenance management of an installation.

Factors to keep in mind for the reckoning of this indicator are two:

- Annual hours.

- Annual hours of stoppage and load reduction exclusively due to the non-programmed corrective maintenance.

As can be seen, in order to reckon this objective, it is not consider neither the hours dedicated to programmed preventive maintenance that imply plant stoppage, nor the hours dedicated to programmed corrective maintenance. For the coherent and correct reckon of this factor, the definition of which is the dissimilarity between programmed and non-programmed corrective maintenance has to be provided always. Therefore, in several industrial installations is common to consider a failure detected, but whose repair could be delayed 48 hours or more, it is considered to be programmed corrective maintenance, and thus it does not count to the reckoning of the reliability.

The objective of maintenance seeks this parameter to always be above an intended value established in the design techno-economical design of the plant, and its value is frequently quite high (same or higher than 99,0%). A well-managed installation ought not to have any problem to reach this value.

The Shelf Life of a Plant

The fifth great maintenance objective is to guarantee a long shelf life for the installation. In other words, the industrial plants must present a condition of deterioration consonant with what intended, in a way that neither the availability nor the reliability nor the maintenance cost could be out of the objectives settled for a long period of time, usually in accordance with the repayment term of the plant. The lifespan of a typical industrial installation is placed between 20 and 30 years, when the benefits of the plant and its maintenance objectives must always be inside the values intended.

A bad-managed maintenance, with a low proportion of hours intended to preventive tasks, with poor budget, lack of means and personnel, and based in provisional repairs, provokes the rapid degradation of any industrial installation. It is remarkable of bad-managed plants that although little time has passed since its initial start, the visual aspect does not correspond to its youth (in terms of lifespan).

The Achieving of the Budget

The objectives of availability, reliability and shelf life cannot be obtained at any cost. The department of maintenance has to achieve the objectives intended adjusting its expenses to what was settled in the annual budget of the plant. As said in the previous section, this budget has to be reckon with extreme care, since a lower budget than the plant needs worsen irretrievably the results of production and limit the shelf life of the installation; on the other hand, an upper budget than the installation needs worsen the operating account results.

Types of Maintenance

When talking or writing about types of maintenance, it is in fact about classifying maintenance tasks into different categories.

Types of Maintenance in Accordance with the Specialization of the Technician

In the first place, maintenance can be divided into the following general types, in accordance with the specialization of the technician who carries out the tasks:

- Operational maintenance, the one carried out by operation personnel.

- Mechanical maintenance, made by mechanic specialists.

- Electrical maintenance, carried out by electric specialists.

- Maintenance of instrumentation, executed by instrumentation specialists.

- Control maintenance, made by control specialists.

- Technical cleanings, carried out by technicians specialized in cleaning certain part of the equipment which require specific and complex procedure.

Types of Maintenance in Accordance With Scope

In accordance with the scope of the tasks, maintenance can be divided into the following types:

- Routine maintenance generally carried out by production personnel.

- Underway maintenance carried out by maintenance staff yet not needing plant, area or a whole system stoppage. Thus only implying a narrow number of equipment and subsystems which do not affect production.

- Minor revisions or inspections, when a limited amount of elements are inspected or substituted.

- Major revisions, when tasks to be accomplished imply the substitution of a great number of wearing elements, or the inspection of certain internal parts that demand great dismantling.

Types of Maintenance in Accordance with Anticipate Faults

Lastly, according to anticipate faults, there are three major groups of maintenance tasks. In several cases, when talking about `types of maintenance´, one maybe thinking about this last category. The types of maintenance according to anticipate faults are the following:

- Corrective maintenance, gets done after the failure has been made and has as its essential goal, its own adjustment.

- Diagnosis, has as its essential goal the knowing of the machines or installation condition in order to decide if an intervention has to be done on it. Generally, an observation or measurement is related to the condition of an installation. Diagnosis tasks frequently include an estimation of its criticality, evaluating the potential severity in cases when the potential and the degradation trend of failure gets to materialize. These type of tasks used to be known as Predictive Maintenance, even though nowadays this name is disused, favoring the concept of Diagnosis, which is a wider concept and which defines better the willfulness of the tasks. The diagnosis includes four types of tasks:

- Easy inspection tasks, very often generally carried out by production technician. It involves simple sensorial inspections, commonly made with the senses, without needing measure tools or additional technical means. Thus including visual inspection, observation of strange noise and vibration emissions and the recognition of abnormal odours. It also includes the reading and record of operating parameters, with instruments installed in the equipment. They require basic training and they can be carried out by any technician. Due to its simplicity these inspections can be done very often.

- Online diagnosis tasks, which are carried out basing on the readings obtained from the inline mounted instrumentation and which are received in the control system.

- Offline diagnosis tasks, which are carried out with instruments that are especially assembled for making the observation or measurements which are used to diagnose. Such examples are the analysis of vibration, thermography, the analysis of ultrasound, oils, fumes produced by combustion and so forth. These are made with offline instruments.

- Detailed inspection of tasks, carried out by specialized maintenance technicians and which may or may not require the stoppage of the equipment and systems to make these inspections. They often demand disassembly or at least, profound observation. As well as this inspections need specific education for the technician that carries them out, who has to be able to distinguish between a standard situation, a situation which requires observation in order to verify its evolution, and an appalling situation which demands immediate intervention. Among these detailed inspections, the followings stand out:

- Mechanic verification, like clearance, alignment, thickness, bolt tightening, starting, operation and stoppage measurements.

- Electrical verification, such as grounding verification, verification of the operation of emergency stop, connexion verification, and so forth.

- Verification of measuring instruments and functional check of control links. They may require an intervention in order to adjust certain parameters to default values.

- Check of certain ways of operation or the benefits of an item.

- Preventive maintenance, carried out before a failure made and which has as its main propose prevent from happening. Preventive maintenance can be further divided into four subtypes, always taking into account that it is actually about subtypes of maintenance inside the category of 'preventive maintenance':

- Conductive maintenance the one carried out by operation technicians and that generally refers to senses check, data samples, change in fuel and/or adjustment of parameters.

- Systematic maintenance, made from time to time or when a certain number of hours has passed.

- Hard-time maintenance, overhaul, major revision or zero hours, which is the combination of tasks made after some time of equipment, system or installation operation, and which has as its goal return the inspected ensemble to its initial state (as when it has zero hours of functioning.

- Improvement maintenance, which is the combination of tasks which are carried out in a part of the installation, and has as goal the avoidance of a certain failure to be made or made again. Although some authors doesn't consider modifications as 'maintenance tasks', it is logic to consider them this way, when they have as goal the avoidance of failures. And they shouldn't be consider 'maintenance tasks' when they only find improvements in security, in the environmental or production impact, without affecting failure probability.

Zero Hours Maintenance (Overhaul)

The set of tasks whose goal is to review the equipment at scheduled intervals before appearing any failure, either when the reliability of the equipment has decreased considerably so it is risky to make forecasts of production capacity.

Preventive Maintenance

Preventive maintenance refers to regular, routine maintenance to help keep equipment up and running, preventing any unplanned downtime and expensive costs from unanticipated equipment failure. It requires careful planning and scheduling of maintenance on equipment before there is an actual problem as well as keeping accurate records of past inspections and servicing reports. Preventive management can be very complex, especially for companies with a lot of equipment. For this reason, many companies rely on preventive maintenance software to help organize and carry out all their preventive maintenance needs.

Preventive maintenance involves the systematic inspection of equipment where potential problems are detected and corrected in order to prevent equipment failure before it happens. In practice, a preventive maintenance schedule may include things such as cleaning, lubrication, oil changes, adjustments, repairs, inspecting and replacing parts, and partial or complete overhauls that are regularly scheduled.

The exact preventive maintenance required will vary based on operation and type of equipment. Recommended standards of the American National Standards Institute (ANSI) are used to help determine the type of inspections and maintenance needed and how often they should be performed. ANSI helps ensure the health and safety of consumers by creating and overseeing the use of thousands of guidelines and norms for nearly every industry, and ANSI standards can be used like a preventive maintenance checklist to define requirements and instructions for maintaining equipment.

Preventive maintenance includes much more than simply performing routine maintenance on equipment. It also involves maintaining accurate records of every inspection and servicing, as well as knowing the lifespan of each part to understand the replacement frequency. These records can help maintenance technicians anticipate the appropriate time to change parts and can also help diagnose problems when they occur. Preventive maintenance software helps collect and organize this information so it is readily available to maintenance technicians.

Working of Preventive Maintenance

Preventive maintenance is a straightforward program to establish and set in motion. Maintenance is set on a schedule based on calendar dates or usage, often at the recommendation of the manufacturer. During a specified date and time, equipment is shut down, and maintenance professionals perform the outlined tasks on that piece of equipment.

Preventive maintenance can also be set up with breakdown and time-based triggers. Maintenance triggers are used to alert employees that maintenance must be performed at an operational level. Breakdown maintenance triggers occur when a piece of equipment breaks down and cannot be used until maintenance is performed. With a time trigger, maintenance is triggered whenever the calendar rolls over to a pre-specified date.

For example, most forklift manufacturers suggest performing preventive maintenance every 150 to 200 hours of operation, which can be established on a time-based trigger. Performing this maintenance can mean extending the life of assets, increasing productivity, improving overall efficiency and reducing maintenance costs.

Preventive maintenance does not require additional tools other than the manufacturer recommendations and a team willing to adopt new maintenance processes. To achieve buy in from the team, it is important to outline the benefits of a PM program, and identify the ways in which a preventive maintenance schedule will make the jobs of technicians, mechanics and engineers much easier.

With a computerized maintenance management system (CMMS), preventive maintenance is triggered for periodic inspections based on calendar intervals or usage (for air compressors and forklifts) or mileage for company vehicles. This company has increased their planned maintenance percentage from 20% to 80%, and their on-time completion rate for PMs is 85% and continues to improve.

Preventive Maintenance Tools

A CMMS is designed to help schedule, plan, manage and track maintenance activities. The features within a CMMS such as task generation, scheduling, inspections and data integration from tools and sensors work together to offer continuous improvement and support for an organization's preventive maintenance program.

- Preventive maintenance task generation: Within a CMMS system, users can leverage a preventive maintenance calendar and/or meter-based PM tasks for all assets and include detailed descriptions with how-to guides and other information vital to effectively performing the work.

- Preventive maintenance task schedules: Developing an effective preventive maintenance program requires more than generating preventive maintenance tasks, and CMMS systems have the tools to make important improvements. Preventive maintenance schedules empower users to coordinate labor resources and parts needed to complete work, as well as automatically generate preventive maintenance tasks based on a daily, weekly or monthly basis, or based on usage.

- Preventive maintenance inspections: A CMMS solution can also help organizations keep up with inspections and pass compliance audits. With CMMS, users can record inspections accurately and generate corrective work orders when equipment fails inspections.

- Tools & sensors: Combined with a CMMS, condition monitoring software enables maintenance managers to remotely monitor multiple assets. It also allows for producing asset alarms and multiple data graphs per asset with correlating current, voltage, temperature, vibration and power quality monitoring.

- Data integration: Data can be integrated into CMMS functionality to enable the completion of seamless workflows on a mobile device. This allows maintenance teams to respond to fault notifications on the move and then create, access or process work orders related to the notification in real time. Planned and unplanned maintenance is better coordinated, unscheduled downtime is reduced and response times to problems or systems failure are improved.

Benefits of Preventive Maintenance

With an effective preventive maintenance program, organizations experience improvements in their overall business processes and costs, including increased productivity, decreased waste, improved work execution and reduced unexpected breakdowns. A preventive maintenance program (with the support of a CMMS and other tools) can help spark serious quantifiable results, such as:

- Extending the life of assets, and increasing equipment up time

- Reducing manual data entry

- Decreasing paperwork with mobile maintenance capability

- Increasing productivity and efficiency

- Improving audit compliance with extensive documentation

Advantages of Preventive Maintenance

Other than reactive maintenance, preventive maintenance is the simplest maintenance strategy to implement and execute, as it requires following manufacturer recommendations and developing a static maintenance schedule for critical equipment. It helps organizations avoid unplanned breakdowns, lost production as well as equipment and labor downtime. It also decreases the cost of maintenance.

Disadvantages of Preventive Maintenance

A common issue that organizations run into with preventive maintenance schedules is performing the inappropriate amount of maintenance on assets. Because calendar-based maintenance does not take the health of an asset into account, the frequency of maintenance work can often be too high or too low. This can be prevented by optimizing and improving preventive maintenance programs.

Tips to Begin Developing a Preventive Maintenance Schedule

1. Establish equipment list and determine best PM candidates: To get started, take note of all of the equipment throughout your organization to establish inventory. Within this list, you will ask yourself the above questions to help decide which pieces of equipment you will include in your future preventive maintenance plan.

2. Refer to manufacturer recommendations: Take a look through manufacturer recommendations to establish an effective preventive maintenance schedule and to help figure out the necessary tasks and desired frequency of maintenance. Referring back to the original example, this could be getting an oil change for your car every 10,000 miles, or as recommended by your owner's manual.

3. Start with your heavy hitters: To effectively leverage a preventive maintenance schedule, it is important to begin with your most critical pieces of equipment one step at a time. Once you get started with those critical assets, create long term plans such as annual schedules.

4. Fill in short term plans: With long term plans established, you can begin creating weekly plans for your crew. These tasks should be assigned and scheduled ahead of time, with all parts and maintenance resources accounted for.

With preventive maintenance scheduling software like CMMS, organizations have experienced benefits such as:

5. Extended asset life and increased equipment uptime.

6. Decreased manual data entry.

7. Less paperwork with mobile and barcoding capability.

8. Consistent maintenance tasks and procedures.

9. Increased productivity and efficiency.

10. Improved audit compliance with extensive documentation.

Applications of Preventive Maintenance

There are many applications of preventive maintenance in a wide variety of industries such as:

- Performing calendar-based maintenance on air conditioning units on a university campus.

- Scheduling consistent maintenance on assets critical for production.

- Meter-based PMs for maintenance of material handling equipment based on utilization.

- Scheduling PM tasks in preparation for audits.

Planned Maintenance

Planned maintenance is a proactive approach to maintenance in which maintenance work is scheduled to take place on a regular basis. The type of work to be done and the frequency varies based on the equipment being maintained, and the environment in which it is operating.

The primary objective of planned maintenance is to maximize equipment performance by keeping equipment running safely for as long as possible, without that equipment deteriorating or having unplanned outages.

Planned maintenance activities include any maintenance work scheduled in advance. For example, changing the oil in a vehicle because the oil light came on is not planned maintenance. Changing the oil because the vehicle had gone 3,000 miles would be planned maintenance. Planned Maintenance is a scheduled maintenance activity, or service visit, that is done to ensure that the equipment, or equipment components, is operating correctly and within the manufacturer's recommendations.

Planned maintenance includes preventive maintenance tasks such as checking oil levels, when those tasks are preplanned.

The schedule for planned maintenance tasks can be based on equipment running hours, number of items produced, distance traveled, or other measurable factors.

Planned Maintenance - Using Computer Systems

Since planned maintenance tasks are done on a regular schedule, they can be used to provide information to feed a computerized system that tracks maintenance needs, as well as being themselves scheduled using a computerized system. However, remember that a computer is only a place to save information and schedule tasks. It does not design your planned maintenance system nor determine how to get the work done.

Working with the production department, and at times a service representative from the manufacturer, equipment maintenance that needs to be regularly scheduled should to be identified and an appropriate schedule developed. A method for measuring the effectiveness of the planned maintenance work should also be designed so that the overall planned maintenance system, as well as the individual tasks, can be evaluated.

One of the key goals of planned maintenance is to ensure the equipment is in compliance with specifications and to proactively identify potential performance problems. The objective is to find and correct problems before they cause an unscheduled outage.

Types of Planned Maintenance

There are many types of planned maintenance that the average workplace performs. Something as simple as general cleaning can be considered a type of maintenance, since removing dirt and debris from an area can help avoid problems with many types of machinery. Some other types of planned maintenance include:

- Lubrication: Most machines require lubricants to keep things running smoothly. Replacing the oil or other lubricant on a regular basis helps avoid many mechanical problems.

- Parts Replacement: Some parts are designed with a set lifespan. A saw blade, for example, will wear out after a set amount of time. If the blade is not replaced on a schedule, it will begin to cut less efficiently. This can also result in cuts that aren't 'clean.'

- Upgrades: Part of many planned maintenance plans is upgrading equipment. Keeping things up to date with the latest options can help improve the safety and efficiently of machines. Planning upgrades on a schedule like this also allows for predictable expenses.

Major Benefits of Planned Maintenance

The key benefit of planned maintenance is that the work can be scheduled in ways such that it does not interfere with overall production. This may mean scheduling planned maintenance during times of the year when there is less demand, or doing maintenance at night when most of the facility is not operating.

Other direct benefits of planned maintenance include:

- Reduces unplanned equipment downtime and improves overall equipment performance.

- Repair costs are reduced because problems are fixed while they are minor.

- More efficient use of manpower and other resources because maintenance work, parts inventory, tools and financial costs, can be spread more evenly throughout the year.

- Better planning of spare parts use and ordering.

- Helping to ensure the manufacturer's requirements for warranty compliance are met.

- Reducing overall maintenance costs.

There are also a number of secondary benefits that come from using planned maintenance. These include:

1. Improved workplace safety.

2. Procedures are established to plan the use of, monitor, and control maintenance resources.

3. Improves the communication between maintenance and operations.

4. Provides a daily plan for maintenance supervisors such that employees have a full day of work every day.

5. Establishes a performance monitoring system that allows maintenance activities to be better evaluated and improved.

Implementing a Planned Maintenance System

As with any other major change, a successful implementation of planned maintenance requires the agreement and participation of everyone affected by the timing and quality of maintenance. This includes the maintenance department, production, safety and possibly the instrument shop. In addition, support from upper level management is needed to ensure the necessary resources to get planned maintenance started are available.

Next get suggestions from, and listen to those who are working with the equipment. They'll be the most knowledgeable about its operation and its problems.

Consider making lean manufacturing techniques such as TPM (Total Productive Maintenance), Kaizen, and 5S part of your planned maintenance system.

Make safety a part of establishing a planned maintenance system. When evaluating equipment to determine the need for planned maintenance, also conduct a Job Hazard Analysis. This will reveal potential hazards that can occur during maintenance as well as bringing out safety issues that can develop during normal operations.

Department Responsibilities

While the production department will have a major voice in scheduling planned maintenance, and if TPM is being used equipment operators may perform some planned maintenance tasks, the bulk of the responsibility for planned maintenance work will fall on the maintenance department. In addition, in cases in which outside expertise is required, either manufacturer's field technicians or qualified third party service personnel will need to be a part of the planned maintenance schedule.

The maintenance department's responsibilities might include:

- Training supervisors so they have the necessary skills and knowledge.

- Maintain a sufficient staffing level, without over-staffing, such that all planned maintenance can be accomplished. Train maintenance workers so they are competent to perform the required tasks.

- Ensure all work is done to the required specifications (including using appropriate fluids and replacement parts).

- Inform the purchasing department about planned maintenance needs with sufficient lead time to allow parts and supplies to be ordered on a normal basis.

- Track maintenance tasks, maintain records of the work that was done, and ensure work is done properly.

- Report on equipment problems noticed during planned maintenance so they can be addressed during turn-arounds or scheduled plant outages.

The above list is only intended to provide some ideas. The staff in each facility is organized differently, and there are different priorities. For example, part of the maintenance department's responsibility might be to minimize staffing through scheduling and judicious use of overtime.

The Role of Labels and Signs in Planned Maintenance

Signs and label play a crucial role in a successful planned maintenance system. For example, every lubrication point should be labeled to identify it and to identify the type of lubrication that should be used. Labels should be used to identify by part number locations where spare parts are used. And of course, labels and signs should be used to warn about hazards and provide safety information.

Corrective Maintenance

Corrective maintenance is any maintenance performed to return equipment to proper working order. Depending on the context of its use it may refer to maintenance due to a breakdown, or maintenance identified through a condition monitoring program.

Corrective maintenance performed due to a breakdown could be either planned or unplanned. In this case, planned corrective maintenance is likely to be the result of a run-to-failure maintenance plan, while unplanned corrective maintenance could be due to an breakdown not stopped by preventative maintenance, or a breakdown due to a lack of a maintenance plan (this is the same as reactive maintenance). Unplanned, maintenance, like reactive maintenance, is much more costly than planned maintenance.

Maintenance performed due to condition-based monitoring software is planned maintenance. It will be scheduled due to a condition trigger. Provided the root cause of the need for maintenance is also addressed, this is an ideal maintenance type.

Advantages of Corrective Maintenance

- Lower short-term costs: As a reactive activity, there is very little to actually do after the purchase and before a problem occurs.

- Minimal planning required: Corrective maintenance consists in correcting a failure identified in a particular component of an equipment or installation reported at the moment, so there's no need for a complex and timely planning.

- Simpler process: The process is easy to understand, since you only need to take action when some kind of problem occurs.

- Best solution in some cases: When it is believed that the stop and repair costs in case of failure will be less than the investment required for preventive maintenance, Corrective maintenance is the best solution.

Disadvantages of Corrective Maintenance

- Unpredictability: The equipment is not monitored after purchase, so the failures are highly unpredictable.

- Paused operations: Unexpected failures may result in unavailable materials and therefore delay the time taken for a repair, increasing equipment downtime.

- Equipment not maximized: This approach doesn't protect or look after equipment and therefore reduces the lifetime of the assets.

- Higher long-term costs: Corrective maintenance is applied when it is believed that the stop and repair costs in case of failure will be less than the investment required for planned maintenance. But this doesn't always happen. When a catastrophic failure occurs, it can be extremely costly, causing negative effects on reputation, client satisfaction, and safety and on the ability to run a business efficiently and productively.

Predictive Maintenance

Predictive maintenance is maintenance that directly monitors the condition and performance of equipment during normal operation to reduce the likelihood of failures. It attempts to keep costs low by reducing the frequency of maintenance tasks, reducing unplanned breakdowns and eliminating unnecessary preventive maintenance.

With predictive maintenance, organizations consistently monitor and test conditions such as lubrication and corrosion. Methods for accomplishing predictive maintenance include infrared testing, acoustic (partial discharge and airborne ultrasonic), vibration analysis, sound level measurements and oil analysis. Computerized maintenance management systems (CMMS), condition monitoring, data integration, and integrated tools and sensors can also facilitate success with condition monitoring.

For example, CMMS empowers companies to define boundaries for acceptable equipment operation, import readings, graph results and automatically trigger an email or generate a work order when boundaries are exceeded.

Working of Predictive Maintenance

Predictive maintenance evaluates the condition of equipment by performing periodic or continuous (online) equipment condition monitoring. Most predictive maintenance is performed while equipment is operating normally to minimize disruption of everyday operations. This maintenance strategy leverages the principles of statistical process control to determine when maintenance tasks will be needed in the future.

For example, rather than changing a vehicle's oil because drives hit 3,000 miles, predictive maintenance empowers organizations to collect oil sample data and change the oil based on the results of asset wear. For predictive maintenance to be effective, it requires both hardware to monitor the equipment and software to generate the corrective work order when a potential problem is detected. Specific types of predictive maintenance include:

- Vibration analysis: Vibration sensors can be used to detect degradation in performance for equipment such pumps and motors.

- Infrared: Infrared cameras are often used to identify unusually high temperature conditions.

- Acoustic analysis: Acoustic analysis is performed with sonic or ultrasonic tests to find gas or liquid leaks.

- Oil analysis: Oil analysis determines asset wear by measuring an asset's number and size of particles.

Additionally, tools such as CMMS, condition monitoring, connected tools and sensors, and data integration can help companies act on the analytics collected by these devices and sensors.

Predictive Maintenance Tools Integrated into a CMMS

Whether you need to track assets through oil viscosity, temperature or vibration, the tools within CMMS systems can help develop accurate predictions when a piece of equipment will require maintenance or replacement.

- Condition Monitoring: Within CMMS systems, condition monitoring tools help empower organizations to execute on predictive maintenance programs. Users can define boundaries of acceptable operation for assets and auto-generate work order or emails when readings fall outside of predefined boundaries.

- Connected sensors & tools: These can offer real-time data streams to track events from anywhere and view AC/DC voltage, current, power and temperature data. By wirelessly syncing measurements taken using handheld tools and comparing them to condition monitoring data, organizations can gain the full picture of equipment efficiency and health.

- Data integration: Data can be integrated into CMMS functionality to enable the completion of seamless workflows on a mobile device. This allows maintenance teams to respond to fault notifications while they are on the move, and then they can create, access or process work orders related to the notification in real time. Planned and unplanned maintenance is better coordinated, unscheduled downtime is reduced and response times to problems or systems failure are improved.

Benefits of Predictive Maintenance

Studies have shown that organizations spend approximately 80% of their time reacting to issues rather than proactively preventing them. Predictive maintenance puts predictive maintenance ahead of the game. It helps predict failures and actively monitor performance. As a result, it saves time and money. Organizations that commit to a predictive maintenance program can expect to see significant improvements in asset reliability and a boost in cost efficiency, such as:

- 10x Return on Investment (ROI)

- 25-30% reduction in maintenance costs

- 70-75% elimination of breakdowns

- 35-45% reduction in downtime

- 20-25% increase in production

The best predictive maintenance programs take time to develop, implement and perfect. The timeline to achieve gains such as these varies, but some clients see positive returns in as little as a year.

Advantages and Disadvantages of Predictive Maintenance

Predictive maintenance requires more time and effort to develop then a preventive maintenance schedule. To be truly effective, employees must be trained on how to use the equipment and interpret the analytics they pull. However, once the commitment is made, predictive maintenance can revitalize not only a maintenance team, but an organization as a whole. There are conditions monitoring contractors who can perform the labor required and analyze the results for your organization.

Applications of Predictive Maintenance

There are many applications of predictive maintenance in a wide variety of industries such as:

- Finding three-phase power imbalances from harmonic distortion, overloads, or degradation or failure of one or more phases.

- Identifying motor amperage spikes or overheating from bad bearings or insulation breakdowns.

- Locating potential overloads in electrical panels.

- Measuring supply side and demand side power at a common coupling point to monitor power consumption.

- Capturing increased temperatures within electrical panels to prevent component failures.

- Detecting a drop-in temperature in a steam pipeline that could indicate a pressure leak.

Implementation of a Predictive Maintenance Strategy

Implementing a predictive maintenance program should be a methodical process from start to finish. The key is to have a long-term view of what to do in order to put all of the foundational components into place.

1. Design the predictive maintenance program: Get positive buy in from management and be prepared to discuss and quantify the benefits and goals. Identify which equipment to target for the program by taking a close look at equipment failure histories and the associated root causes. Equipment that is failing the most will provide the most potential for cost reductions and reliability improvements. Compare the cost of implementing a predictive maintenance to the average cost of equipment failures. As stated above, sometimes predictive maintenance does not make sense. Depending on the asset, a corrective method of maintenance could be cheaper.

2. Select predictive maintenance technology: Choose which of the above technologies would be most effective to monitor the condition of your equipment. Is your organization more interested in vibration analysis, infrared thermography, ultrasonic inspection or oil analysis? Select the tools that will provide that information.

3. Allocate proper resources: Develop and train an implementation team to perform predictive maintenance activities. Carve out time in the schedule for predictive maintenance tasks such as data collection, analysis, reporting and tracking, and allocate funding for predictive maintenance technology investments or, for a predictive maintenance, contractor to assist.

4. Perform system integrations: Leverage the tools within and integrated into a CMMS to help turn condition monitoring data into action. For example, a company offering equipment monitoring services, lubrication engineering and reliability engineering can record negative diagnostic reports and automatically generate corrective work orders.

5. Coordinate preventive maintenance & predictive maintenance programs: Leveraging both preventive and predictive maintenance makes for the best maintenance programs. Use each method where applicable and decide which strategy to apply based on disruption due to equipment downtime, cost of parts and labor time, and equipment history.

6. Utilize CMMS reports & dashboards: With reporting and dashboard tools, organizations can consistently document work order history, failures, costs and trends. This helps to track progress for key stakeholders.

Reliability Centered Maintenance

RCM or Reliability Centered Maintenance is one technique more within the possibilities to develop a maintenance plan in an industrial plant that it has some important advantages over other techniques. Initially, it has been developed for the aviation sector, where the high costs derive from the systematic replacement of parts threatened the airlines profitability. Later, this technique has been transferred to the industrial field, after establishing the excellent results it has had in the aeronautical field.

RCM was firstly documented in a report written by FS Nowlan and H.F. Heap and published by the Department of Defense of United States in 1978. Since then, the RCM has been used to help to formulate strategies for managing physical assets in virtually all areas of organized human activity, and in virtually all industrialized countries in the world. This process defined by Nowlan and Heap has been the basis for various application documents in which the RCM process has been developed and refined in subsequent years. Many of these documents retain key elements of the original process. However, the widespread use of the name "RCM" has led to the emergence of a large number of failure analysis methodologies that differ significantly from the original, but its authors also called "RCM". Many of these other processes fail to achieve the objectives of Nowlan and Heap, and some are even counterproductive. In general try to abbreviate and summarize the process, leading in some cases completely denature.

As a result of international demand for a rule setting minimum standards for a failure analysis process can be called "RCM" emerged in 1999, the SAE JA 1011 standard and in 2002 the SAE JA 1012 standard. These rules not intend to be a guide or manual procedures, but merely set out, as noted, criteria that should satisfy.

Total Productive Maintenance

TPM is a maintenance philosophy aimed at eliminating production losses due to equipment status, or in other words, keeping equipment in a position to produce at maximum capacity, the expected quality products, with no unscheduled stops. This includes:

- Zero breakdowns

- Zero downtimes

- Zero failures attributed to poor condition of equipment

- No loss of efficiency or production capacity due to this equipment

It is understood perfectly the name: total productive maintenance, or maintenance that provides maximum or total productivity.

The Eternal Fight Between Maintenance and Production

Maintenance has traditionally been seen with a separate and external part to the production process. TPM emerged as a need to integrate the maintenance department into the operation or production one to improve productivity and availability. In a company that TPM has been implemented, all the organization works on maintaining and improving equipment. It is based on five principles:

- All staff participation, from senior management to plant operators. Including every one of them can guarantee the success of the objective.

- Corporate culture creation oriented to obtain maximum efficiency in the production system and management of equipment and machinery. 'Global efficiency' is pursued.

- Management system implementation of production plants to facilitate the elimination of losses before they happen.

- Preventive maintenance implementation as an essential mean to achieve the objective of zero losses through integrating activities in small groups and it is based on the support provided by the autonomous maintenance.

- Management systems implementation of for all aspects of production, including design and development, sales and management.

Six Great Losses

Since the philosophy of TPM, it is considered that a machine stop for making a change; a breakdown in a machine; machine which does not work at 100% of capacity or that manufactures defective products; are intolerable situations which cause losses to the company. The machine should be considered unproductive in all these cases, and appropriate actions designed to avoid them in the future should be taken. TPM identifies six sources of loss (called the 'six great losses') that reduce the effectiveness because of interfering with the production:

- Equipment failure, producing unexpected loss of time.

- Commissioning and machine settings (or downtime), which produce loss of time to start a new operation or another stage of it. For example, in the early morning, when changing workplace, when changing a mould, or when making an adjustment.

- Idle, waits and minor stoppages (minor failures) during normal operation which cause loss of time, either by problems in the instrumentation, small obstructions, etc..

- Reduction of operation speed (the machine does not operate at full capacity), which causes production losses because the design speed of the process is not achieved.

- Defects in the process, production losses happen because we need to remake parts of it, reprocess defective products or complete unfinished activities.

- Loss of time typical from commissioning of a new process, idle, probationary period, etc..

Careful analysis of each of these causes of low productivity leads to finding solutions to eliminate them and the means to implement these last ones. It is essential that the analysis is done by the production staff and maintenance staff together, because the problems that cause low productivity are from both types and solutions should be adopted in an integral way to their success.

Participation of The operator in Maintenance Tasks

From a practical point of view, implementing TPM in an organization means that maintenance is perfectly integrated into production. Therefore, part of the maintenance work have been transferred to production staff, who no longer feels the equipment like something that others take care of it, but as their own, they have to pamper and repair it, the operator feels the equipment as his.

Maintenance involves differences in three levels:

- Operator's level, who will be responsible for operative maintenance tasks that are very simple, such as cleaning, adjustment, parameters monitoring and minor failures repairing.

- Integrated technical level. Within the production team, there is at least one maintenance person who works with the production staff; he/she is just one of them. This person solves problems more significance, for which more knowledge is needed. But he is there, close, so you do not have to tell anyone or wait. The spare part is also decentralized: each production line, even each machine, have everything it takes closely.

- For higher-level interventions, and scheduled overhauls that involve complex disassemblies, delicate adjustments, etc., the company has a maintenance department which is not integrated into the structure of production. But it handles common tools.

Operator's involvement in maintenance tasks makes him better understand the machine and the installation he is operating, its characteristics and capabilities, its criticality. Also, it helps to work as a team, and it facilitates sharing of experiences and mutual learning, and this improves personal motivation.

There is a fundamental difference between philosophy of TPM and RCM: while the TPC is based on the people and the organization as the center of the process, the RCM maintenance is based on failure analysis, and preventive measures to be taken to avoid them, rather than on people.

Maintenance Models

Each of the models presented below include several of the previous types of maintenance at the indicated rate. Moreover, all of them include two activities: visual inspections and lubrication. This is because it is demonstrated that these tasks realization in any equipment is profitable. Even in the simplest model (Corrective Model), in which virtually the equipment is left on its own and we do not deal with it until a fault occurs. It is advisable to observe it at least once a month, lubricate it with suitable products to their characteristics. Visual inspections virtually no cost money (these inspections will be included in a range where we have to look at other nearby equipment, so it will not mean we have to allocate resources specifically for this function). This inspection allows us to detect faults in an early stage and its resolution will generally be cheaper as soon as detected. Lubrication is always profitable. Although it does represent a cost (lubricant and labor), it is generally so low that it is more than warranted, since a malfunction due to a lack of lubrication will always involve a greater expense than the corresponding to lubricant application.

With this remark, we can already define the various possible maintenance models.

Corrective Model

This is the most basic model, and includes, in addition to visual inspections and lubrication mentioned previously, the arising breakdowns repair. It is applied, as we will see, to equipments with the lowest level of criticality, whose faults are not a problem, economically or technically. In this type of equipment is not profitable to devote more resources and efforts.

Conditional Model

It includes the activities of the previous model, and also this model carries out a series of tests that will determine a subsequent action. If after testing we discovered an anomaly, we will schedule an intervention; on the contrary, if everything is correct, we will not act on the equipment.

This maintenance model is valid in equipment not to very used, or for equipment that despite being important in the production system the probability of failure is low.

Systematic Model

This model includes a set of tasks we will perform no matter what is the condition of the equipment , also we will perform some measurements and tests to decide whether to carry out other tasks of greater magnitude, and finally, we will repair faults that arise. It is a model widely used in equipment of medium availability, of some importance in the production system whose failures cause some disruption. It is important to note that equipment subjected to a systematic maintenance model does not have to have all its tasks with a fixed schedule. Just a equipment with this model of maintenance can have systematic tasks that are carried out regardless of the time it have been operated or state of the elements on which it works. It is the main difference with the previous two models in which to perform a maintenance task should be some sign of failure.

An example of equipment subjected to this maintenance model is a discontinuous reactor, in which the tasks that must react are introduced at once, the reaction takes place, and then the reaction product is extracted before making a new load. Regardless of this reactor is doubled or not, when operating should be reliable, so it is warranted a series of tasks regardless of whether any signs of failure have been arose.

Other Examples

- The landing gear of an aircraft

- The engine of an aircraft

High Availability Maintenance Model

It is the most demanding and exhaustive model of them. It is applied to that equipment that under no circumstances may suffer a breakdown or malfunction. These are equipments to whom are also required very high levels of availability, above 90%. The reason for such high level of availability is generally high cost in production due to a fault. With a demand so high, there is no time to stop the equipment if the maintenance requires it (corrective, preventive, systematic).

To maintain this equipment is necessary to use predictive maintenance techniques that allow us to know the status of the equipment when is working, and scheduled shutdowns, which supposes a complete overhaul, with a frequency usually annually or higher.

Examples of this model of maintenance may be:

- Turbine of power production.

- high temperature furnaces, where an intervention means cooling and re-heating the furnace, resulting in energy expense and production losses associated with it.

- Rotating equipment working continuously.

- Reactor deposits or reaction tanks not duplicated that are the basis of production and to be kept in operation as many hours as possible.

Other Considerations

When designing the Maintenance Plan should be taken into account two important considerations affecting some equipment in particular. Firstly, some equipment are subjected to legal rules that regulate their maintenance, forcing them to perform certain activities with an established frequency.

Secondly, some of the maintenance activities cannot be performed with the regular maintenance equipment (either their own or hired) because it requires knowledge and / or specific resources that are only up to the manufacturer, distributor or a specialist team.

These two aspects should be assessed when trying to determine the maintenance model that we should apply to an equipment.

Legal Maintenance

Some equipment is subjected to rules or regulations by the Administration. Above all, there are equipment that are hazardous to people or the environment. The Administration requires the completion of a series of tasks, tests and inspections, and some of them must be performed by companies duly authorized to carry them out. These tasks must necessarily be incorporated into the Maintenance Plan of the equipment, whatever model you decide to apply.

Some of the equipment subjected to this type of maintenance are:

- Equipment and devices under pressure

- Installation of High and Medium Voltage

- Cooling Towers

- Certain lifts: service or people

- Vehicles

- Fire Prevention Facilities

- Storage tanks of certain chemicals

Subcontracted Maintenance to a Specialist

When we talk about a specialist, we refer to an individual or a company specialized in a particular equipment. The specialist may be the equipment manufacturer, importer's technical service, or a company that has specialized in a particular type of intervention. As we said, we must turn to a specialist when:

- We do not have sufficient knowledge

- We do not have the necessary resources

If there are these circumstances, some or all of maintenance work must be outsourced to specialized companies.

The subcontracted maintenance to a specialist is usually the most expensive alternative, as the company offering it is aware that not compete. The prices are not market prices, but monopoly prices. You should try to avoid it as far as possible by cost increase and higher external dependence that it involves. The most reasonable way to avoid this is to develop a training plan that includes specific training for those equipment that do not have enough knowledge also acquired the necessary technical means.

Maintenance Engineering

Maintenance Engineering focuses on determining which maintenance tasks are necessary and which working methods should be used. Maintenance engineering is also about developing a maintenance strategy and a maintenance plan. A successful maintenance strategy is usually based on the corporate strategy of an organization.

The aim of a maintenance plan that is crafted by our maintenance engineers is to reduce risks and increase the availability of machines. The result of a successful maintenance plan is higher revenues for your entire organization.

The role of Maintenance Engineer, also called a Plant Engineer, covers everything from equipment and component maintenance strategy selection, equipment maintenance cost modelling, life cycle analysis, operational risk management—right through to re-engineering equipment for reliability improvement.

A Maintenance Engineer is the second person called-in when equipment is not operating properly. The first person called to address problem plant and equipment is the appropriate Maintenance Technician. When the technician cannot solve the problem the maintenance engineer is called in to help resolve the issue correctly. Since the job involves solving difficult engineering problems a Maintenance Engineer needs to understand the engineering design of their equipment and the science of their processes. It requires a person with the scientific knowledge to analyze the physics of a situation and the mathematically capability to model and calculate the engineering and process dynamics occurring.

To be a great Maintenance Engineer you must be expert in how your plant and equipment work, how they are designed and constructed, and how they are correctly used. There is no better way to be successful as a Maintenance Engineer than to know why problems arise in your plant and know how to properly solve them. This is especially the case when it becomes necessary to improve the reliability of plant and equipment. To make plant highly reliable you must extend their life between outages by twice and three times; hopefully many times more. To make that much difference to your operation you must know and understand the causes of failure and then find excellent answers that greatly extend lifetimes. To do that you must understand the applicable engineering, process science, and process control and materials-of-construction properties so well that you confidently make good choices.

Similarly when you need to select maintenance strategy, choose maintenance activities and set maintenance frequencies for plant and equipment, you will make far better decisions when you know the engineering and the operating properties and parameters of your process and equipment.

Typical Maintenance Engineering Responsibilities

Typical responsibilities include:

- Assure optimization of the Maintenance Organization structure.

- Analysis of repetitive equipment failures.

- Estimation of maintenance costs and evaluation of alternatives.

- Forecasting of spare parts.

- Assessing the needs for equipment replacements and establish replacement programs when due.

- Application of scheduling and project management principles to replacement programs.

- Assessing required maintenance tools and skills required for efficient maintenance of equipment.

- Assessing required skills for maintenance personnel.

- Reviewing personnel transfers to and from maintenance organizations.

- Assessing and reporting safety hazards associated with maintenance of equipment.

Maintenance Management

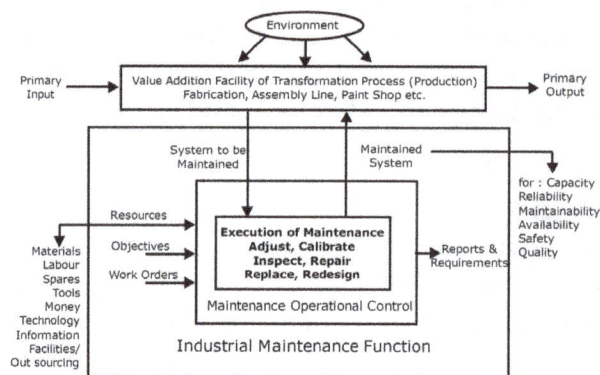

Figure: "An integrated input-output model for maintenance function".

Maintenance management is the direction and organization of resources in order to control the availability and performance of industrial plant to some specified level (Gillett, 2001). Maintenance is a function in an organization that operates in parallel with production. Moreover besides being a support function, it has a role in gaining and maintaining competitive advantages. Therefore, it is very important for all relevant stakeholders to be aware of the role of maintenance in achieving sustainable and competitive business environment. An integrated model for maintenance function has been conceptualized and depicted in figure 2. The primary output of production is the desired product while demand for maintenance would be the secondary output as a result of production activities. This output would act as input for the maintenance function. Maintenance results in restored production capacity which would further act as secondary input to production thus completing the maintenance cycle. Thus production manufactures the product while maintenance produces the capacity for production. The quality of the final product is affected by both the production process and the quality of maintenance work.

Thorsteinsson and Hage proposed a broad definition of the maintenance task based on viewing the maintenance system as a "Production System" where the "Products" are maintenance services.

Maintenance Testing

Once a system is deployed it is in service for years and decades. During this time the system and its operational environment is often corrected, changed or extended. Testing that is provided during this phase is called maintenance testing.

Usually maintenance testing is consisting of two parts:

- First one is, testing the changes that has been made because of the correction in the system or if the system is extended or because of some additional features added to it.

- Second one is regression tests to prove that the rest of the system has not been affected by the maintenance work.

Importance of Testing

1. To ensure the reliability and efficiency of the equipment and/or facilities.

2. To provide smooth and satisfactory operation and performance within the facility without compromising the safety.

3. To determine critical areas or equipment that needs immediate action.

4. As the saying goes prevention is better than cure, by doing scheduled maintenance and inspection, all issues concerning the equipment or facilities can be taken into action immediately.

Machine Element

Design activity involves endowing form and ensuring functionality of a given product. The function of many products depends on devices that modify force or motion and consist of a number of interrelated units. An example is a gear box that modifies the torque speed characteristic of a prime mover to match that of the load, and comprises gears, shafts, bearings, and seals. Such devices that modify force or motion are called machines, and the interrelated units are sometimes known as machine elements. The technology base has developed into a mature status in this area, providing a number of such devices that can be used, in some cases, as is or can be designed as fit-for-purpose.

In the design of a machine and its machine elements, it is the designer's task to determine the motion, forces, and changes in energy involved so that sizes, shapes, and materials for each machine element making up the whole machine can be determined. Although design work may involve concentrating on one component at a given time, it is essential to consider its interrelationship with the whole product, following the total design philosophy and taking into account the market requirement, detailed specifications, the conceptual design, the design of other components, and the manufacturing requirements.

The goal in machine design is to determine the size, shape, and selection of materials and manufacturing processes for each of the parts of the machine, so that the machine will perform its intended function without failure. This requires the designer to model and predict the performance and operation of each component and overall assemblies, studying the mode of failure for each and ensuring that conditions expected in service for the product are met. This may involve determining stresses, deflections, temperatures, and material degradation for each and every component and overall assembly, as well as its influence within the total design. As stresses are a function of forces, moments, and torques, consideration of the dynamics of the system will often be necessary before a detailed stress analysis can be undertaken. Furthermore, as the strength of materials is a function of temperature, a thermal analysis may also be necessary in order to determine operating temperatures throughout the performance cycle of the machine.

If a machine has no moving parts, then the tasks for the designer are simpler because only static force analysis and thermal analysis are necessary. However, structures are subject to loading from external loads such as weather conditions of wind and rain, earth tremors, and traffic. As such, consideration of the environment into which the machine will be installed is warranted and appropriate failure modes mitigated against. Mitigation involves considering what may go wrong and determining what action to takedto either prevent it from going wrong in the first place or, if it does go wrong, what is going to be done about it.

If the motions involved in a machine are slow then it is likely that a static force analysis will be sufficient. In a static structure, such as the floor of a casing designed to support a particular weight, the payload that the structure can support can be increased by adding appropriately distributed weight (dead weight) to its structure. If, however, the components of a machine have significant accelerations, a dynamic force analysis will be necessary. Adding additional mass to a moving component increases the forces involved, and it is possible for such components to become victims of their own mass. This is because some of the loading causing stress in a moving component is due to inertial forces modeled by Newton's second law of motion, embodied in the following equation:

$$F = ma$$

Although the addition of mass may increase the strength of a component, the benefit may be reduced by the increases in inertial forces.

To meet a design requirement, the designer needs to conceive a form and express the connectivity of its parts in order to meet the desired goal. A proportion of the mental sculpting involved is geometric in nature because shape often permits the achievement of a given function. The design procedures presented here follow this approach, with an initial proposal for the geometric configuration followed by analysis to justify the proposal or provide guidance for its modification. Boden discusses creativity in terms of the connection of conceptual spaces in the mind. In order to create a design, it is often necessary to combine ideas from different fields or disciplines. This may or may not take place as part of a structured design procedure or in a prescribed time period. Over-reliance on procedure should, in some cases, actually be guarded against, as this may stymie the activity.

The variety of machine elements and technologies already available to the designer as a result of years of development is extensive, Machine elements are those more commonly found in machines transmitting power by means of moving components, such as motors and engines. The diagram given in figure illustrates the range of machine elements involved in the design of a gear box for a

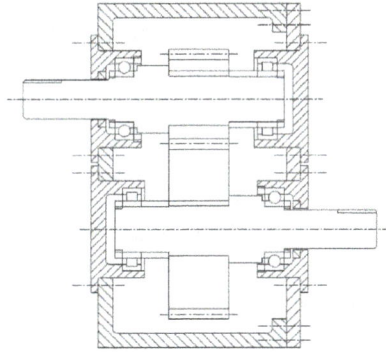

single cylinder air compressor and comprises shafts, bearings, gears, seals, and enclosure. The machine elements considered about the design of such applications and cover bearings, shafts, gears, belts and chains, seals, clutches and brakes, springs, fasteners, and enclosures.

A fundamental element for many machines is the lever, and the principle of leverage is used in many machine elements as well as machines. The fixed point of the lever, about which it moves, is known as the fulcrum. A lever can be used with a fulcrum (also known as a pivot) to allow a small force moving over a large distance to create a large force moving over a short distance. The positions of the force and the load are interchangeable; by moving them to different points on the lever, different effects can be produced. There are three generic classes of levers: first-order lever, second-order lever, and third-order levers. An example of a first- order lever is a seesaw, or balance, with the load and the force separated by the fulcrum. As one side moves up, the other side moves down. The amount and the strength of the movement are proportional to the distance from the fulcrum. If L_2 is $10 \times L_1$, then F_1 will be $10 \times F2$.

$$F_1 = F_2 \frac{L_2}{L_1}$$

An example of a second-order lever is a wheel barrow. Here, the load is between the force and the fulcrum. This uses mechanical advantage to ease lifting of a large weight. In a third-order lever, the force is between the fulcrum and the load. Mechanical advantage is reduced, but the movement at the load point is increased. An example of a third-order lever is a tweezer.

A further fundamental building block in machine design and many mechanisms is the linkage. A linkage is a system of links, rods, or spars connected at joints with rotary or linear bearings. Linkages can be used to change direction, alter speed, and vary the timing of moving parts. An example of the use of linkages is shown in figure where the articulation of the grab bucket is enabled by a four-bar linkage, giving prescribed rotation in response to the hydraulic ram actuation. A four-bar linkage is also used to provide mechanical advantage for grasping a work piece in mole grips.

In the example shown in figure, two linked linkages are used to convert the small linear movement of the drive shaft (bottom left); first into a rotational body movement and second into a fast hammer movement. In this case, the speed of the input drive shaft is amplified by the use of the linkages providing high-speed hammer action.

Many machine elements involve relative motion between components and associated friction and wear. Tribology is the study of friction, lubrication, and wear of surfaces in relative motion. Because of the desire to produce efficient machines, where the loss of energy due to friction is minimal, extensive research and effort have been put into studying tribology. As a subject, it has considerable interrelationships and cross-over with materials and fluid.

Multiple uses of linkages in construction equipment, JCB JS210.

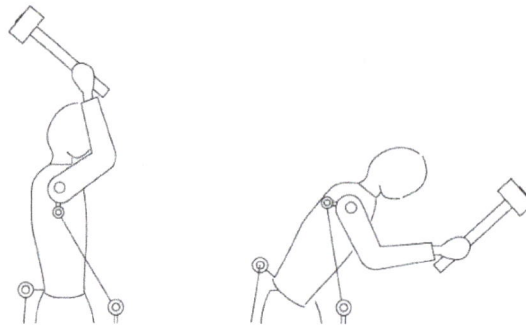

Figure: Two linked linkages.

Bearings

The term bearing, in its general sense, refers to the assembly formed by two surfaces that has the capacity for relative motion. A wide range of bearings has been developed; some involve rolling motion, sliding motion, and both rolling and sliding motions. Lubrication of a bearing is frequently used to reduce friction and, therefore, the wear and power absorbed by the bearing, as well as to remove heat. This ensures the operation of the bearing assembly at temperatures compatible with the materials and lubricant used.

Typical examples of bearings are shown in Figures below illustrates rolling bearings. These come in many forms, each designed for a given set of performance characteristics, such as speed and load capability. Rolling element bearings usually need to be lubricated by grease or oil with the choice depending upon the speed and temperature of operation. Rolling element bearings are typically available as stock items. This is of significant potential advantage to a machine designer, as the task becomes that of bearing selection rather than bearing design. Bearing design is a significant undertaking, involving consideration of very high point contact stresses and use of special materials. The technology involved relies heavily on the science of tribology and there are a number of worldwide companies already operating in this market. A designer needs to carefully consider the issue of whether time is better spent either on producing a new design for a bearing that is already available from an existing supplier or on designing and adding value to the product in hand.

Figure: Examples of rolling element bearings available from existing suppliers.

Figure: Fixed geometry journal (multilobe) bearings.

Plain bearings rely on either rubbing contact between the bearing surfaces or pressure to separate the two surfaces. The pressure required can be induced in a lubricant by the movement of one component relative to another or by the external supply of the lubricant under the required pressure. Plain bearings are also available from specialist bearing suppliers. In principle, they can be significantly less expensive than rolling element bearings because they involve fewer moving components. The selection of an appropriate bearing for a given task, however, is an involved activity, which needs to take into account, among other factors:

- Load,

- Speed,

- Location,

- Size,

- Cost,

- Starting torque,

- Noise, and

- Lubrication supply.

A consequence of bearings being supplied by a select number of specialist manufacturers is that they have been subject to standardization and are usually only available in a limited number of standard sizes. Standardization has an advantage in that the parts from one original equipment manufacturer (OEM) will be interchangeable with another in terms of their geometry. This does not, however, mean that the performance of two similarly sized components from different manufacturers will be the same.

Gears, Belts, and Chains

Gears, belts, and chains can be used to transmit power from one shaft to another. Typical examples are illustrated in Figures below.

Figure: Spur gears incorporated within an epicyclic gear box for a cordless hand-tool transmission.

Many power producing machines, or prime movers, such as internal combustion engines, gas turbines, and electric motors, produce power in the form of rotary motion. The operating torque versus speed characteristics of prime movers vary according to their type and size, as do the driven loads. It is common for the torque versus speed characteristics to be mismatched, requiring the need for gearing and, perhaps also, a clutch to enable the prime mover to attain sufficient momentum prior to engaging a load.

When transmitting power from a source to the required point of application, a series of devices are available including belts, pulleys, chains, hydraulic and electrical systems, and gears. Generally, if the distances of power transmission are large, gears are not suitable and chains and belts can be considered. However, when a compact, efficient, or high-speed drive is required, gear trains offer a competitive and suitable solution.

Figure: Flat belt and synchronous belt drives driving multiple devices from a single belt.

Figure: Chain drive for powering multiple elements.

For example, for a small electric motor-driven air compressor, options that might be possible for the transmission include: belt drive; chain drive; gear drive; belt and chain; belt and gear; and gear and chain. Three different types of machine element drives along with direct drives are all potential solutions that could be viable. The different concepts are illustrated as general arrangements in Figure below.

Gears, belts, and chains are all available from specialist suppliers. The manufacture of gears requires specialist machinery, and the stresses in the region of contact between gears in mesh can be exceptionally high requiring the use of specialist materials. Similarly, the design of roller chain and associated sprockets involves high contact stresses and the design of an articulated joint and associated bearings. For these reasons, combined with financial expediency, leaving the cost of specialist manufacture infrastructure to other companies, gear, chain, and belt manufacture is usually best left to the specialist. As such, the task of the machine designer becomes that of selecting and specifying appropriate components from the specialist stock suppliers and original equipment manufacturers.

Clutches and Brakes

Clutches and brakes provide frictional, magnetic, hydraulic, or mechanical connection between two machine elements, usually shafts. There are significant similarities between clutches and brakes. If both shafts rotate, then the machine element will be classed as a clutch; the usual function is to connect or disconnect a driven load from a driving shaft. If one shaft rotates and the other is stationary, then the machine element is classed as a brake; the likely function is to decelerate a shaft. In reality, however, the same device can function as a brake or clutch by fixing its output element to either a shaft that can rotate or to ground, respectively.

Brakes and clutches are familiar devices from their use in automotive applications. They are, however, also used extensively in production machinery. Clutches allow a high inertia load to be rotated with a smaller electric motor than would otherwise be required if it was directly connected. Clutches are also used to maintain a constant torque on a shaft for tensioning of webs and filaments. They can be used, in emergencies, to disconnect a driven machine from a motor if machinery should become jammed. In such cases, a brake may also be installed in order to bring the machinery to a rapid stop.

Complete clutches and brakes are available from specialist manufacturers. In addition, key components such as disks and hydraulic and pneumatic actuators are also available, enabling the designer to opt, if appropriate, to design the overall configuration of the clutch or brake and buy in the frictional lining surface.

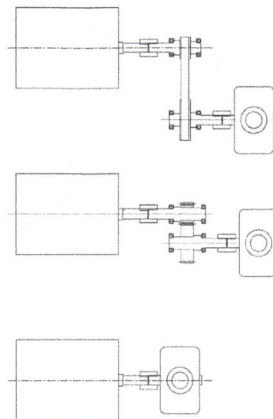

Figure: Conceptual arrangements for a transmission.

Seals

Seals are devices used to reduce or eliminate the flow of fluid or particulates between two locations. Seals are an important part of a machine design where fluids, under relatively high differential pressures, must be contained within a particular region. An example is the fluid within the cylinder of an internal combustion engine. Here, the working fluid needs to be contained in the cylinder for part of the cycle and excluded, in as much as possible, from contaminating the lubricating oil for the crankshaft bearings. This is conventionally achieved by means of piston rings, figure.

Figure: Piston ring seals. There are many types of seals, but they can generally be classified as either static or dynamic depending on whether relative movement is involved. Many static seals involve the use of a gasket or sealing material, such as an elastomer. These are widely available from specialist suppliers, and the task facing the designer is the selection of appropriate materials for the application. Some dynamic seals are also commercially available, such as radial lip seals and mechanical face seals. However, there is not the level of standardization in shaft design. In addition, the range of possible applications is seemingly infinite with differing pressure ratios and fluids being sealed. As a result, the range of seal geometries required is too extensive to warrant an original equipment manufacturer designing or stocking a wide number of seal sizes and types. Instead, many dynamic seals are designed fit-for-purpose. Typical considerations in the selection and design of seals include the nature of the fluid or particulates to be contained, the differential pressure, the nature of any relative motion between components, the level of sealing required, operating conditions, life expectancy, and total costs.

Springs

Springs are flexible elements used to exert force or torque and store energy. The force exerted by a spring can be a linear push or pull, or a radial force similar to that of a rubber band around a roll of paper. Alternatively, a spring can be configured to produce a torque with applications including door closers. Springs store energy when they are deformed and return energy when the force that is causing the deflection is removed. A further type of spring is the power spring, or spring motor, which once wound up can dissipate energy at a steady pace.

Springs can be made from round or rectangular wire bent into a suitable form such as a coil, or made from flat stock and loaded as a cantilevered beam. Many standard size spring configurations are available as stock items from specialist manufacturers and suppliers, and it is usually expedient to use these where possible. It is important to focus on both the performance characteristics of the spring and the spring's ends or terminations. Many options are available such as twist loops or hooks, side loops, extended hooks, etc. Some applications require a custom spring design. Spring design involves consideration of many variables such as length, wire diameter, forces, spring rate, spring index, number of coils, pitch, pitch angle, installation considerations, coil clearance, materials, types of loading, and allowable stresses.

Fasteners

Fasteners connect or join two or more components. It has been estimated that there are over 2.5 million fasteners in a Boeing jumbo jet. For more modern airliners, such as the Airbus Industry A340, this amount of fasteners has been dramatically reduced, but the total is still significant. This example illustrates the importance of fasteners to the designer.

There are thousands of types of fasteners commercially available from specialist suppliers. The most common include threaded fasteners such as bolts, screws, nuts, studs and set screws, rivets, fast operating fasteners requiring a quarter turn to connect or disconnect, and snap joints commonly used in plastic component assembly. In addition, there is the range of permanent fastening techniques by means of welding, brazing, soldering, and using adhesives.

Given the wide range of fastening techniques available and the costs involved in sourcing, stocking, and assembly, the correct choice of fastener is very important.

Wire Rope

Wire rope is sometimes used in hoisting, haulage, and conveyer applications. By using many small diameter wires, twisted around a central core, some flexibility in the wire rope can be achieved. It is possible for the wire rope to articulate drums and other radial segments.

Pneumatics and Hydraulics

Significant mechanical advantage can be achieved by exposing a surface area to a pressurized fluid. This principle is exploited in many hydraulic and pneumatic devices, which have been used extensively in machine design and are available as standard stock items.

Enclosures

Individual machine elements such as shafts on bearings must be supported, and this is the function of an enclosure or frame. The total design model requires attention to all other phases of the design process while undertaking any one task. As such, the design of any given machine element should not occur in isolation; but instead give due consideration to how it will fit in and interact with the overall assembly. The enclosure for a machine provides both the support for the machine and the form to a product. Visual attractiveness is not always given the attention it warrants by designers of machine elements; however, we produce our components to be purchased. Current thinking indicates that visual perception is dominant among the senses, with over one-half the cerebral cortex dedicated to visual processing. Therefore, the shapes of our products are important, and we should design accordingly.

Enclosures are usually unique to a given class of machine, as each machine tends to be different in terms of the number, size, and function of its constituent components. It is not practical, therefore, to have a completely general approach to enclosure and frame design. Common examples of the variations of frames and enclosures are those for vehicles and toys. Toys must be safe, functional, and fun; yet the material and personnel effort in producing the toy should minimize the production costs and maximize profits.

Figure: Bus frame design.

The design of an enclosure or frame is an art to some extent, in that the constituent machine components must be accommodated. There may be constraints on the location of supports in a machine such that they facilitate the function and operation of the machine and one can access particular components. The frame for the bus shown in figure: illustrates this point in that access for doors, windows, and occupants must be provided.

Some of the more important design parameters to consider for an enclosure or frame include: strength, size, assembly, appearance, corrosion resistance, stiffness, vibration, weight, noise, production costs, maintenance costs, and sustainability. A list such as this may be too all- encompassing to be of much use at the start of a design process. The following list of factors may be more helpful to consider as the onset for a frame or enclosure:

- Forces exerted by the machine components through mounting points such as bearings, pivots, brackets, and feet.

- Requirements to support the frame itself.

- Allowable deflections of constituent machine elements.

- Cooling requirements.

- Environments into which the machine will be transported and installed in relationship to other machines and infrastructure.

- Quantity required.

- Production facilities available.

- Expertise available for the design task. The complexity involved when considering more than one machine element is evident. Due to the need to "keep the bigger picture in mind," models for the design process such as total design are useful.

References

- Mobley, Keith R. & Higgins, Lindley R. & Wikoff, Darrin J. (2008) Maintenance Engineering Handbook, Mc-Graw-Hill Professional, Seventh Edition, 2008, ISBN 0-07-154646-4, ISBN 978-0-07-154646-1

- What-is-maintenance: petrochemicalmaintenance.com, Retrieved 11 May 2018

- What-is-preventive-maintenance: micromain.com, Retrieved 31 March 2018

- Dhillon, Balbir S. (2006) Maintainability, Maintenance, and Reliability for Engineers, CRC Press, 2006, ISBN 0-8493-7243-7, ISBN 978-0-8493-7243-8

- Planned-maintenance: graphicproducts.com, Retrieved 18 April 2018

- Corrective-maintenance: fiixsoftware.com, Retrieved 09 July 2018

- Advantages-and-disadvantages-corrective-maintenance: home.infraspeak.com, Retrieved 28 March 2018

- Types of maintenance: mantenimientopetroquimica.com, Retrieved 17 April 2018

Industrial Failures

It is vital to ensure the appropriate functioning and operation of machinery and manufacturing processes for continued production. All the vital aspects of industrial failures that may occur in the operation of industrial plants have been carefully analyzed in this chapter, such as failure analysis, failure cause, fault reporting, failure rate, failure mode, effects, and criticality analysis.

Industrial failure is an Event in which any part of an equipment or machine does not perform according to its operational specifications. Failures are classified into several categories: dependent failure, non-critical failure, random failure, etc.

Main Causes of Industrial Failure

When a piece of machinery fails it inevitably costs a company resources, time and money. Unfortunately machinery can fail for numerous reasons and pinpointing the cause is not always easy. Perhaps a Siraflex spring balancer has gone, a Sirem pump has failed or a Zurrer driver has ceased up? Whilst some causes of machine failure can be obscure and hard to detect, others are much more common.

Take a look at the following main causes of industrial machinery failure.

Accidents

A piece of machinery being handled or operated in an incorrect manner can lead to internal parts becoming damaged and causing failure.

Equipment being dropped can also cause components such as Spaggiari gearboxes and Planox clutches to become dislodged or damaged, which would also cause the machinery to cease working properly.

Inadequate Maintenance

Inadequate maintenance or a lack of maintenance can result in accidents occurring with machinery and equipment ultimately breaking down.

The Agency continues that sufficient maintenance of machinery includes, inspection, testing, measurement, replacement, adjustment, repair, upkeep, fault detection, replacement of parts, servicing, lubrication and cleaning.

Corrosion

Corrosion of vital industrial parts, such as Tschan couplings and Worm gearboxes, is one of the most common causes of equipment failing.

Corrosion of components can be especially problematic when the machinery is exposed to water contamination. As well as creating rust to form on the interior and exterior of the machine, water increases the speed at which oil oxidises, which ultimately leads to the part operating within an acidic environment.

Misalignment

According to Lifetume Reliability's paper titled 'Principle Causes of Failure in Machinery', "misalignment is universally recognized as the leading contributor to machinery failure."

Principally due to the parasitic axial thrust, misalignment can significantly reduce the life of a bearing. Misalignment, continues the paper, exists when the center lines of two neighbouring machines deviate from each other.

Bearing Failure

Which brings us on to bearings. From Amt linear bearings to sferax linear bearings, if bearings fail so too can the equipment you are operating.

Premature bearing failure is most commonly caused by the contamination or loss of bearing lubricant. Although mechanical defects such as the unbalance of misalignment of bearings can also lead to bearings failing prematurely.

Metal Fatigue

Another leading reason industrial machinery can fail and stop working is due to metal fatigue. Metal fatigue occurs when you attempt to cut wires without the use of tools. As the operator makes their way through the wire, the metal works harder to harder, creating fatigue.

When several cycles of such stress have occurred, the metal typically becomes brittle and snaps off.

To help prevent machinery failure and the stress and loss of earnings equipment breakdown typically creates, it is important to keep machinery and internal parts well lubricated, well maintained, have parts regularly replaced, keep it stored in an appropriate place and only operated by those trained to do so.

Failure Analysis

Failure analysis is the systematic investigation of a part failure with the objectives of determining the root causes of failure and the corrective actions needed to prevent future failures. Failures occur when some system or part of a system fails to perform up to the expectations for which it was created.

Failure analysis is commonly performed in the following industries:

- Chemical processing

- Refining

- Oil & gas

- Pulp & paper

Failure analysis relies on collecting failed components for subsequent examination of the cause or causes of failure using a wide array of methods, especially microscopy and spectroscopy. Key benefits of failure analysis include avoiding costly future instances of corrosion-induced failure, and providing evidence for any litigation proceedings.

Analysis of a failed part can be done using destructive testing or non-destructive testing (NDT). The NDT methods (such as industrial computed tomography scanning) are valuable because the failed products are unaffected by analysis. Therefore, these methods are normally used before any others.

Generally, procedure for failure analysis includes:

1. Collection of data and samples

2. Preliminary examination

3. Non-destructive inspection

4. Mechanical testing

5. Selection and preservation of damaged surfaces

6. Macroscopic and microscopic examination

7. Preparation and examination of metallographic sections

8. Damage classification

9. Report writing

Determining the root cause of a failure is a three-part process.

Step 1: Data Collection

The first step in a root cause failure analysis is data collection. We will also work with you to determine your goals for the failure analysis examination, determine how the part should operate, and consult with additional subject matter experts, if needed.

The type and breadth of questions we ask during this step can be surprising to an organization that has not performed failure analysis previously. Some of the information important to collect during this phase includes:

- What are the steps of the manufacturing process?

- What is involved in each process step?

- What are all of the components in the device?

- What are the specific component's purposes?

- What are the ways in which each of these components could fail?

- How severe is the impact of these potential failures?

- What types of device failures are actually occurring?

- At what point does a failure affect the device's function?

- What is the likelihood device failures are noticed before they reach the consumer?

- What is the history of the failed unit?

- What is the occurrence of failures?

During this phase, we will conduct tests on the product being analyzed. It is common to test a representative sample of failed devices as well as those that are working properly. This can help to determine what components are failing and when the failure occurs.

Whenever possible, we prefer non-destructive tests in the failure analysis testing. This tends to provide the most accurate data and is more economically efficient for the customer. Sometimes the failure analysis requires cross-sections of the material or performs thermal testing. In these cases, we perform these tests later in the data collection process.

Step 2: Analyze Data Collected to Determine Root Cause Failure

The next step in the failure analysis process is to determine the root cause of the failure. Device failure is rarely the result of a single incident. Our experience has shown there are multiple inputs into even a "simple" failure. This indicates there are often different ways to prevent the failure in the future.

Consider this example of an electronic device failure.

Many of the recommendations for correcting a problem are small changes that can have a significant impact. Small changes in how source materials and product components are tested, treated and stored can significantly reduce device failure.

In other cases, we find that failure occurs when customers use a device for too long or in the wrong operating conditions. In these cases, our recommendations might include additional product education for marketing and sales staff. These employees can then provide this information to customers. This prevents device failures from improper use or application.

Electronic Component Failure Analysis

Electronic components and hardware failure can occur during many phases of a product's lifecycle. Along with problems during the product design and manufacturing stage, electronic components can fail because of issues with:

- Storage

- Packaging

- Installation

- Operation

- Maintenance

We use both non-destructive and destructive tests to determine the root cause of electronic component failures.

Electronic Device Failure Analysis

Determining the root cause of electronic device failure is often more difficult than determining root cause failure for other objects. Interactions between software and hardware, it's important to find a laboratory that is skilled in electronic device failure analysis.

Metal Failure Analysis

Metal failure can have a big impact on products across the supply chain. From contamination and corrosion that causes medical equipment to fail to stress failures that affect structural integrity, metal failure can have major consequences. It's estimated metal corrosion alone accounts for nearly $300 billion in economic losses each year.

Failure Analysis is a complex process that relies on a variety of techniques.

Plastic/Composite Failure Analysis

Plastics can fail in many different ways. Plastic products can fail from stress fractures, fatigue, material degradation and contamination. Plastic products can also fail in less-serious ways: discoloration and distortion can both affect the integrity of plastic products. Determining the exact cause of plastic failure requires a range of tests and a broad knowledge of polymers.

Root cause testing for plastic failures follows a similar process to that used for metal failure analysis. Plastic failure analysis can be more complex, because plastics often contain additives like plasticizers, colorants and reinforcing fillers. As a result, failure testing for plastic products often requires specialized testing of the molecular and chemical structures in plastics.

Contamination Analysis

If there are contaminants in the supply chain, it is essential to control them quickly. However, identifying contaminants can be a time-consuming process that requires major resources.

Types of contamination analysis typically conducted include:

- Identifying the type and source of foreign particles in a product.

- Determining the level at which a contaminant affects product integrity.

- Tracing the source of contamination and developing corrective plans.

Failure Cause

Rolling element bearings are among the most important and popular components in the vast majority of machines. Additionally, the component most likely to cause machine downtime is the bearing, because all machine forces are transmitted through the bearings. Therefore, rolling element bearings have been the subject of extensive research over the years to improve their reliability. However, since a large number of bearings are associated with any critical process, system failure due to any individual bearing failing can occur in a short period of time. There are many reasons for

Causes of Industrial Bearing Failures

Types of the Bearing Failures and its Causes

Rolling element bearings are among the most important and popular components in the vast majority of machines. Additionally, the component most likely to cause machine downtime is the bearing, because all machine forces are transmitted through the bearings. Therefore, rolling element bearings have been the subject of extensive research over the years to improve their reliability. However, since a large number of bearings are associated with any critical process, system failure due to any individual bearing failing can occur in a short period of time. There are many reasons for early failure, such as heavy loading, inadequate lubrication, careless handling, ineffective sealing, or insufficient internal bearing clearance due to tight fits. Each of these factors results in its own particular type of damage and leaves its own special imprint on the bearing.

Rolling bearing damage may result in a complete failure of the rolling bearing at least, however, in a reduction in operating efficiency of the bearing arrangement. Only if operating and environmental conditions as well as the details of the bearing arrangement (bearing surrounding parts, lubrication, sealing) are completely in tune, can the bearing arrangement operate efficiently. Bearing damage does not always originate from the bearing alone. Damage due to bearing defects in material or workmanship is exceptional. The types of mechanical bearing failure and their frequencies are categorized in Table. The most frequent bearing failure category is corrosion, which is lubrication related. Chemical reaction occurs between the oil and the surface of the bearing, generally from water or other corrosive materials present in the oil. Dimensional discrepancies of rolling element bearings are a consequence of damage prior to or during service. The causes of dimensional discrepancies could be manufacturing flaws, improper handling or installation, and severe overloading during service. Foreign objects, carried by contaminated lubricant, are trapped inside the bearing between the rolling element and the raceway, and are overloaded. Understanding the underlying reason for the defects and their consequences in terms of failures gives the diagnostic clues to detect early failures.

Reason	Failure Percentage
Corrosion	35 %
Dimensional Discrepancies	29 %
Foreign Objects	24 %
Other	10 %

Table: Percentage distribution of the bearing failure

Bearing failures that are not responsible for material fatigue are generally classified as premature. Typical reasons for rolling bearing damage are:

1. Inexpert Mounting:

- Incorrect mounting method, wrong tools
- Contamination
- Too tight fit
- Too loose fit
- Misalignment

2. Abnormal Conditions During operation:

- Overload, absence of load
- Vibrations
- Excessive speeds

3. Unfavorable Environmental Influences:

- External heat
- Dust, dirt
- Of electric current
- Humidity
- Media

4. Inadequate Lubrication:

- Unsuitable lubricant
- Lack of lubricant
- Over lubrication

Each of the different causes of bearing failure generates its own characteristic damage. Such damage is also known as primary damage, which, in turn, creates secondary, failure inducing damages, such as spalling and cracks. Most failed bearings frequently display a combination of primary and secondary damage. The types of damage are summarized above.

Bearing Failure Modes

The normal service life of a rolling element bearing rotating under load is determined by material fatigue and Wear at the running surfaces. Premature bearing failures can be caused by a large number of factors, the common of which are fatigue, Wear, corrosion, brinelling and poor lubrication. Each of the different causes of bearing failure produces its own characteristic damage. Such damage, known as primary damage, gives rise to secondary, failure-inducing damage - flaking and cracks. Even the primary damage may necessitate scrapping the bearings on account of excessive internal clearance, vibration, noise, and so on. A failed bearing frequently displays a combination of primary and secondary damage. The types of damage may be classified Primary damage includes Wear, Indentations, Smearing, Surface distress; Corrosion & Secondary damage includes Flaking, Cracks.

Fatigue

A bearing subject to alternative normal loads could fail due to material fatigue after a certain operation time. Fatigue damage begins with the formation of minute cracks below the bearing surface. As loading continues, the cracks progress to the surface where they cause material to break loose in the contact areas. The actual failure can manifest itself as pitting, spalling or flaking of the bearing races or rolling elements. If the bearing continues in service, the damage will spread in the vicinity of the defect due to stress concentration. The surface damage severely disturbs the nominal motion of the rolling elements by introducing short time impacts repeated at the characteristic rolling element defect frequencies. If the bearing were to continue in service, the damage may spread to other raceways or rolling elements and eventually lead to increased friction and temperature followed by complete seizure.

Wear

Wear is another common cause of bearing failure. It is caused mainly by dirt and foreign particles entered in the bearing through sealing or due to contaminated lubricant. The abrasive foreign particles roughen the contacting surfaces giving a dull appearance. Severe wear changes the raceway profile and increases the bearing clearance. The rolling friction increases considerably and can lead to high level of slip and skidding. The end result of this is complete breakdown. Increasing wear will gradually introduce geometric errors in the bearing. Non-uniform diameters of weared rolling elements will cause cage frequency vibration.

Geometric err of the raceways will result in the production of multiple harmonics of shaft speed being produced. In normal cases there is no appreciable wear in rolling bearings. Wear may, however, occur as a result of the ingress of foreign particles into the bearing or when the lubrication is unsatisfactory. Vibration in bearings which are not running also gives rise to wear. Wear caused by abrasive particles Small, abrasive particles, such as grit that have entered the bearing by some means or other, cause wear of raceways, rolling elements and cage. It is easy to feel where the dividing line goes between wom and unworn sections.

Corrosion

Corrosion damage occurs when water, acids or other contaminants in the oil enter the bearing assembly. This can be caused by damaged seals, acidic lubricants or condensation which occurs when bearings are suddenly cooled from a higher operating temperature in very humid- air. The rust particles also have an abrasive effect which generates wear. The rust pits also form the initiation sites for subsequent flaking and spalling. Rust will form if water or corrosive agents reach the inside of the bearing in such quantities that the lubricant cannot provide protection for the steel surfaces. This process will soon lead to deep seated rust. Another type of corrosion is fretting corrosion.

Deep Seated Rust

A thin protective oxide film forms on clean steel surfaces exposed to air. However, this film is not impenetrable and if water or corrosive elements make contact with the steel surfaces, patches of etching will form. This development soon leads to deep seated rust. Deep seated rust is a great danger to bearings since it can initiate flaking and cracks. Acid liquids corrode the steel quickly, while alkaline solutions are less dangerous. The salts that are present in fresh water constitute, together with the water, an electrolyte which causes galvanic corrosion known as water etching. Salt water, such as sea water, is therefore highly dangerous to bearings.

Fretting Corrosion

If the thin oxide film is penetrated, oxidation will proceed deeper into the material. An instance of this is the corrosion that occurs when there is relative movement between bearing ring and shaft or housing, on account of the fit being too loose. This type of damage is called fretting corrosion and may be relatively deep in places. The relative movement may also cause small particles of material to become detached from the surface. These particles oxidize quickly when exposed to the oxygen in the atmosphere. As a result of the flatting corrosion, the bearing rings may not be evenly supported and this has a detrimental effect on the load distribution in the bearings. Rusted areas also act as fracture notches.

Brinelling

Brinelling, manifest itself as regularly spaced indentations distributed over the entire raceway circumference, corresponding approximately in shape to the Hertzian contact area. Three possible scenarios causing Brinelling are (1) when a bearing is subjected to static overloading which leads to plastic deformation of the raceways, (2) when a stationary rolling bearing is subjected to vibration and shock loads and (3) when a bearing forms a loop for the passage of electric current. In all cases, the result will be repetitive indentations of the raceways. In some instances, a large number of indentations may occur as the bearing may occasionally be turned slightly. The bearing operation will be noisy and uneven in the presence of Brinelling with each indentation acting like a small fatigue site producing sharp impacts with the passage of the rolling elements. Continued operation will lead to the development of spalling at the indentation sites.

Lubrication Starvation

Inadequate lubrication, either in terms of quantity or quality, is one of the common causes of premature bearing failure as it leads to skidding, slippage and bearing seizure. At the highly stressed region of Hertzian contact, when there is insufficient lubricant, the contacting surfaces will weld together. The three critical points of bearing lubrication occur at the cage-roller interface, the roller-race interface and the cage-nice interface.

Indentations

Raceways and rolling elements may become dented if the mounting pressure is applied to the wrong ring, so that it passes through the rolling elements, or if the bearing is subjected to abnormal loading while not running. Foreign particles in the bearing also cause indentations. Foreign particles, such as burrs, which have gained entry into the bearing cause indentations when rolled into the raceways by the rolling elements. The particles producing the indentations need not even be hard. Thin pieces of paper and thread from cotton waste and cloth used for drying may be mentioned as instances of this. Indentations caused by these particles are in most cases small and distributed all over the raceways.

Smearing

When two inadequately lubricated surfaces slide against each other under load, material is transferred from one surface to the other. This is known as smearing and the surfaces concerned become scored, with a ragged appearance. When smearing occurs, the material is generally heated to such temperatures that rehardening takes place. This produces localized stress concentrations that may cause cracking or flaking. In rolling bearings, sliding primarily occurs at the roller end-guide flange interfaces. Smearing may also arise when the rollers are subjected to severe acceleration on their entry into the load zone. If the bearing rings rotate relative to the shaft or housing, this may also cause smearing in the bore and on the outside surface and ring faces. In thrust ball bearings, smearing may occur if the load is too light in relation to the speed of rotation. In certain circumstances, smearing may occur on the surface of rollers and in raceways of spherical and cylindrical roller bearings. This is caused by roller rotation being retarded in the unloaded zone, where the rollers are not driven by the rings. Consequently material is heated to such temperatures that the hardening takes place. This produces localized stress concentrations that may cause cracking or flaking or transfer of material (commonly known as Smearing) from one surface to the other.

Surface Distress

If the lubricant film between raceways and rolling elements becomes too thin, the peaks of the surface asperities will momentarily come into contact with each other. Small cracks then form in the surfaces and this are known as surface distress. These cracks must not be confused with the fatigue cracks that originate beneath the surface and lead to flaking. The surface distress cracks are microscopically small and increase very gradually to such a size that they interfere with the smooth running of the bearing. These cracks may, however, hasten the formation of sub-surface fatigue cracks and thus shorten the life of the bearing. If the lubrication remains satisfactory throughout, i.e. the lubricant film does not become too thin because of lubricant starvation or viscosity changes induced by the rising temperature or on account of excessive loading, there is no risk of surface distress.

Flaking

(Spalling) Flaking occurs as a result of normal fatigue, i.e. bearing has reached the end of its normal life span. However, this is not the common. If cause of bearing failure. The flaking detected in bearings can generally be attributed to other factors. If the flaking is discovered at an early stage, when the damage is not too extensive, it is frequently possible to diagnose its cause and take the requisite action to prevent a recurrence of the trouble. The path pattern of the bearing may prove to be useful. When flaking has proceeded to a certain stage, it makes its presence known in the form of noise and vibrations, which serve as a warning that it is time to change the bearing. The causes of premature flaking may be heavier external loading than had been anticipated, preloading on account of incorrect fits or excessive drive-up on a tapered seating, oval distortion owing to shaft or housing seating out-of-roundness, axial compression, for instance as a result of thermal expansion. Flaking may also be caused by other types of damage, such as indentations, deep seated rust, electric current damage or smearing. Heavy loading and inadequate lubrication are the causes of this damage.

Cracks

Cracks may form in bearing rings for various reasons. The most common cause is rough treatment when the bearings are being mounted or dismounted. Hammer blows, applied direct against the ring or via a hardened chisel, may cause fine cracks to form, with the result that pieces of the ring break off when the bearing is put into service. Excessive drive up on a tapered seating or sleeve is another cause of ring cracking. The tensile stresses, arising in the rings as a result of the excessive drive-up, produce cracks when the bearing is put into operation. The same result may be obtained when bearings are heated and then mounted on shafts manufactured to the wrong tolerances. The smearing described in an earlier section may also produce cracks at right angles to the direction of slide. Cracks of this kind produce fractures right across the rings. Flaking, that has occurred for some reason or other, acts as a fracture notch and may lead to cracking of the bearing ring. The same applies to fretting corrosion.

Causes of Gear Failure

Gear failure can occur in various modes. If care is taken during the design stage itself to prevent each of these failure a sound gear design can be evolved. The gear failure is explained by means of flow diagram in figure.

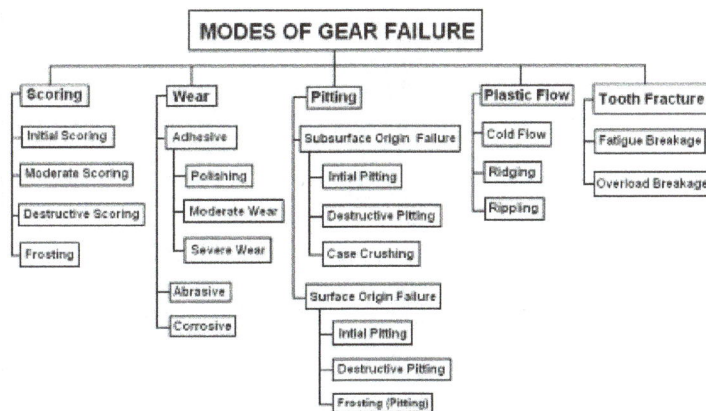

Figure: Different modes of failure

Scoring

Scoring is due to combination of two distinct activities: First, lubrication failure in the contact region and second, establishment of metal to metal contact. Later on, welding and tearing action resulting from metallic contact removes the metal rapidly and continuously so far the load, speed and oil temperature remain at the same level. The scoring is classified into initial, moderate and destructive.

Initial Scoring

Initial scoring occurs at the high spots left by previous machining. Lubrication failure at these spots leads to initial scoring or scuffing as shown in figure below Once these high spots are removed, the stress comes down as the load is distributed over a larger area. The scoring will then stop if the load, speed and temperature of oil remain unchanged or reduced. Initial scoring is nonprogressive and has corrective action associated with it.

Figure: Initial scoring

Moderate Scoring

After initial scoring if the load, speed or oil temperature increases, the scoring will spread over to a larger area. The Scoring progresses at tolerable rate. This is called moderate scoring as shown in figure below.

Figure: Moderate scoring

Destructive Scoring

After the initial scoring, if the load, speed or oil temperature increases appreciably, then severe scoring sets in with heavy metal torn regions spreading quickly throughout as shown in figure below. Scoring is normally predominant over the pitch line region since elastohydrodynamic lubrication is the least at that region. In dry running surfaces may seize.

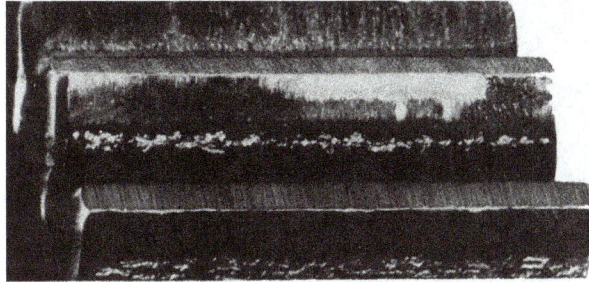

Figure: Destructive scoring

Wear

As per gear engineer's point of view, the wear is a kind of tooth damage where in layers of metal are removed more or less uniformly from the surface. It is nothing but progressive removal of metal from the surface. Consequently tooth thins down and gets weakened. Three most common causes of gear tooth wear are metal-to-metal contact due to lack of oil film, ingress of abrasive particles in the oil and chemical wear due to the composition of oil and its additives. Wear is classified as adhesive, abrasive and chemical wear.

Adhesive Wear – Mild/Polishing Wear

Unlike scoring, adhesive wear is hard to detect. It occurs right from the start. Since the rate of wear is very low, it may take millions of cycles for noticeable wear. Prior to full load transmission, gears are run in at various fractions of full load for several cycles.

Polishing Wear

The surface peaks are quashed over a long period of running and the surface gets polished appearance. Hence this is known as mild or polishing wear as shown in figure above.

Moderate Adhesive Wear

When the load and speed of operation are more than mild wear conditions, moderate wear takes place with higher rate. Worn out portions appear bright and shiny. Yet it occurs over a long period. A typical example of this wear in helical gear is shown in the figure above.

Moderate adhesive wear

Abrasive Wear

Abrasive wear is the principal reason for the failure of open gearing and closed gearing of machinery operating in media polluted by abrasive materials. Examples are mining machinery; cement mills; road laying, building construction, agricultural and transportation machinery, and certain other machines. In all these cases, depending on the size, shape and concentration of the abrasives, the wear will change. Abrasive wear is classified as mild and severe.

Mild Abrasion

Mild abrasion is noticed when there is ingress of fine dust particles in lubricating oil which are abrasive in nature. Since abrasive is very fine, the rate of metal removal is slow. It takes a long time for perceptible wear.

Mild Abrasion

The surface appears as though it is polished. A spiral bevel pinion with mild abrasion is shown in figure above. Mild abrasive wear is faced in cement mills, ore grinding mills. Fine dust particles entering the lubricating medium cause three body abrasions. The prior machining marks disappear and surface appears highly polished as shown in figure below. Noticeable wear occurs only over a long time. Sealing improvement and slight pressurization of the gear box with air can reduce the entry of dust particles and decrease this wear.

Mild Abrasion

Severe Abrasion

This wear occurs due to ingress of larger abrasive particles in the lubricating medium and higher concentration of the particles. The particles will plough a series of groove on the surface in the direction of sliding on the gear tooth as seen in the figure below. High rate of wear in this case will quickly reduce the tooth thickness. Thinned tooth may later on fracture leading to total failure.

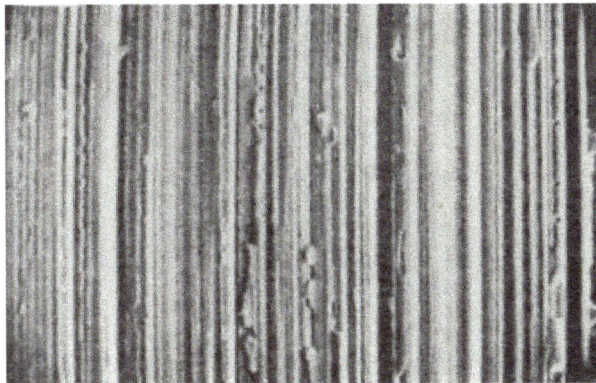

Severe Abrasion

Corrosive Wear

Corrosive wear is due to the chemical action of the lubricating oil or the additives. Tooth is roughened due to wear and can be seen in the figure (a). Chemical wear of flank of internal gear caused by acidic lubricant is shown in figure (b).

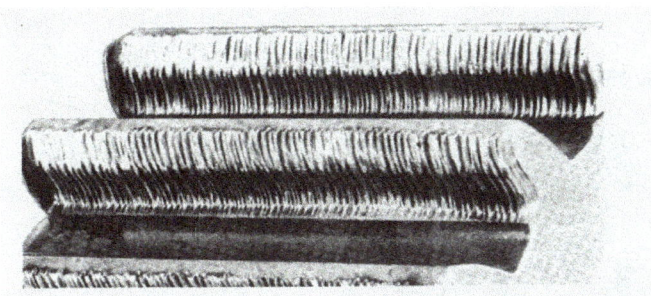
(a)

(b)

Corrosive Wear

Pitting of Gears

Pitting is a surface fatigue failure of the gear tooth. It occurs due to repeated loading of tooth surface and the contact stress exceeding the surface fatigue strength of the material. Material in the fatigue region gets removed and a pit is formed. The pit itself will cause stress concentration and soon the pitting spreads to adjacent region till the whole surface is covered. Subsequently, higher impact load resulting from pitting may cause fracture of already weakened tooth. However, the failure process takes place over millions of cycles of running. There are two types of pitting, initial and progressive.

Initial/Incipient Pitting

Initial pitting occurs during running-in period wherein oversized peaks on the surface get dislodged and small pits of 25 to 50 μm deep are formed just below pitch line region. Later on, the load gets distributed over a larger surface area and the stress comes down which may stop the progress of pitting.

Initial Pitting

In the helical gear shown in figure above pitting started as a local overload due to slight misalignment and progressed across the tooth in the dedendum portion to mid face. Here, the pitting stopped and the pitted surfaces began to polish up and burnish over. This phenomenon is common with medium hard gears. On gears of materials that run in well, pitting may cease after running in, and it has practically no effect on the performance of the drive since the pits that are formed gradually become smoothed over from the rolling action. The initial pitting is non-progressive.

Progressive or Destructive Pitting

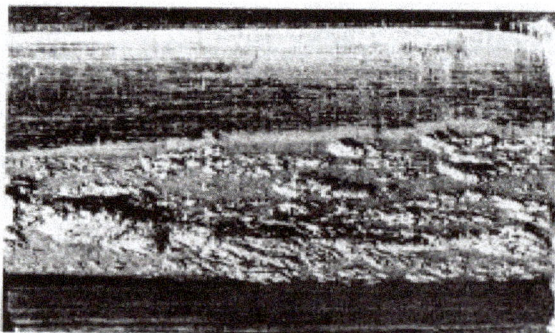

Tooth Surface Destroyed by Extensive Pitting

During initial pitting, if the loads are high and the corrective action of initial pitting is unable to suppress the pitting progress, then destructive pitting sets in. Pitting spreads all over the tooth length. Pitting leads to higher pressure on the unpitted surface, squeezing the lubricant into the pits and finally to seizing of surfaces.

Pitting begins on the tooth flanks near the line along the tooth passing through the pitch point where there are high friction forces due to the low sliding velocity. Then it spreads to the whole surface of the flank. Tooth faces are subjected to pitting only in rare cases. Figure above shows how in destructive pitting, pitting has spread over the whole tooth and weakened tooth has fractured at the tip leading to total failure.

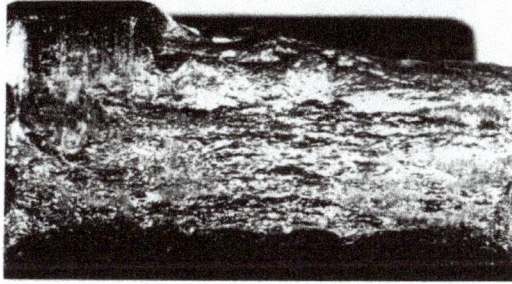

Whole Tooth is Destroyed by Extensive Pitting

Flaking/Spalling

In surface-hardened gears, the variable stresses in the underlying layer may lead to surface fatigue and result in flaking (spalling) of material from the surface as shown in figure below.

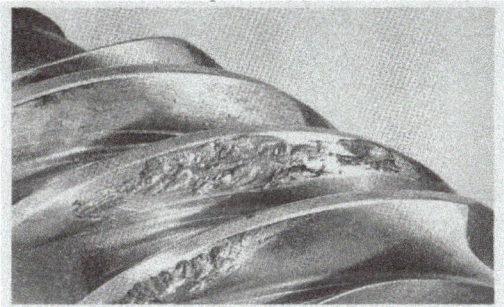
(a)

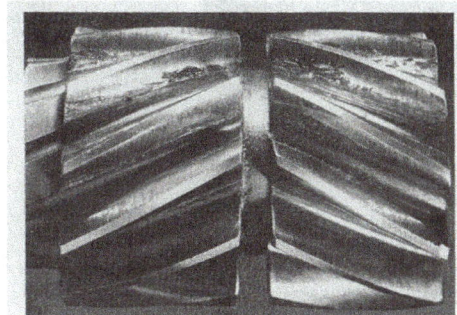
(b)

Figure: Flaking / Spalling

Pitting - Subsurface Origin Failure

Figure below shows the subsurface origin failure.

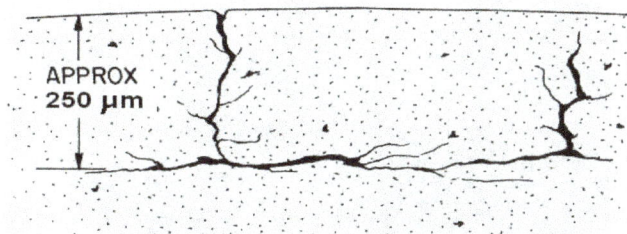

Subsurface origin failure

Pitting - Surface Origin Failure

Failure modes in gear namely the surface origin failure is shown in figure below.

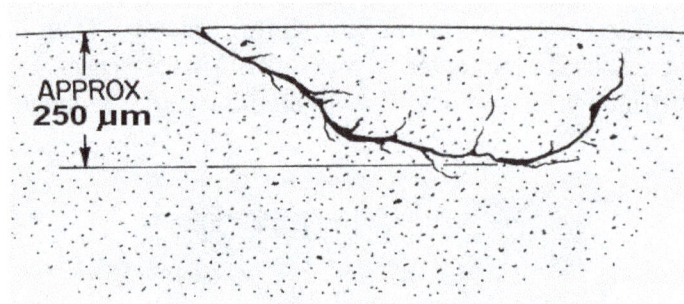

Surface origin failure

Prograssive Pitting

The progressive pitting is shown in figure below.

  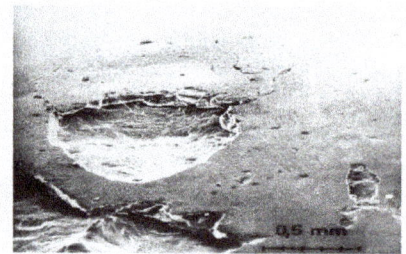

(a)   **(b)**   **(c)**

Figure: Progressive pitting

Pitting - Frosting

Frosting usually occurs in dedendum portion of the driving gear first and later on the addendum as shown in figure below. The wear pattern doesn't have normal metal polish but has etched-like finish.

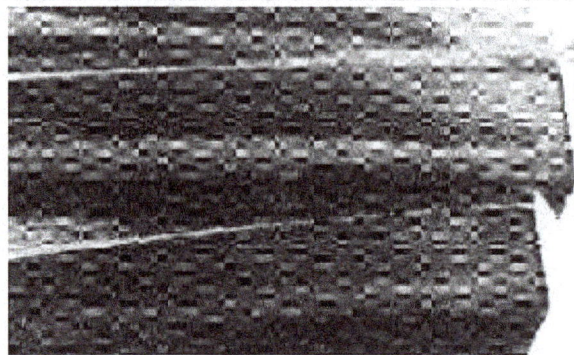

Figure: Frosting

Under magnification, surface reveals very fine micro-pits of 2.5µm deep. These patterns follow the higher ridges caused by cutter marks. Frosting results from very thin oil film and some asperity contact.

Pitting Failure

Surface endurance strength determines the selection of dimensions and material for almost all gearing operating under conditions of the best possible lubrication.

Plastic Flow – Cold Flow

Plastic flow of tooth surface results when it is subjected to high contact stress under rolling cum sliding action. Surface deformation takes place due to yielding of surface or subsurface material. Normally it occurs in softer gear materials. But it can occur even in heavily loaded case hardened gears. Cold flow material over the tooth tip can be seen clearly in the bevel gear shown in the figure below.

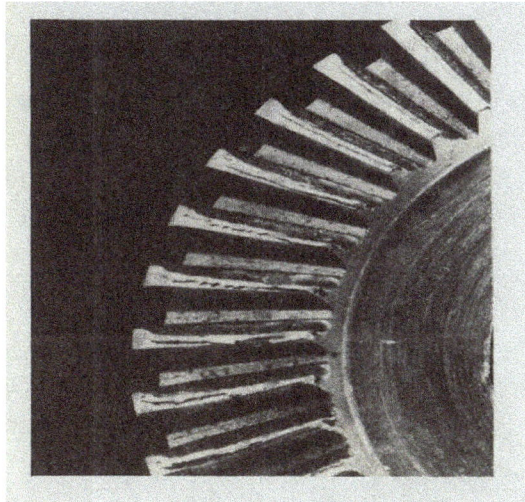

Plastic flow - cold flow

Plastic Flow Due to Overheating

The plastic flow due to overheating is shown in figure below.

Damage to a case hardened gear resulting from overheating
associated with insufficient lubrication

Plastic Flow – Ridging/Grooving

When moderately loaded softer gears run for sometime, they develop a narrow band of bright finish along the pitch line. It is due to reversal of direction of sliding at the pitch line. After running for a longer time or with heavier load, the pair of gears in ductile steel will often exhibit ridge along the pitch line of wheel and groove in the pitch line of the pinion as shown in figure below.

Plastic flow – Ridging / Grooving

In accordance with the direction of the friction forces, the plastic flow of the material on the teeth of driven gear is toward the pitch point and leads to the formation of a ridge at the line passing through this point. In driving gears the flow is away from this line and a groove is formed. Too low viscosity of the lubricant and lack of surface hardness are the reasons.

Rippling in Spiral Bevel Pinion

Rippling is a periodic wave-like formation at right angles to the direction of sliding. It has fish scale appearance and is usually seen on hardened gear surface. It is a kind of wear or plastic deformation in micro form with very thin oil films. Such a failure in a Spiral bevel pinion is shown in the figure below.

Rippling in spiral bevel pinion

Tooth Fracture

Tooth fracture is the most dangerous kind of gear failure and leads to disablement of the drive and frequently to damage of other components (shafts, bearings, etc.) by pieces of the broken teeth. Tooth breakage may be the result of high overloads of either impact or static in nature, repeated overloads causing low-cycle fatigue, or multiple repeated loads leading to high cycle fatigue of the material.

Tooth Breakage – Bending Fatigue

Bending fatigue failure occurs over a long period of time. The initiation of crack takes place at the weakest point, normally at the root of the tooth or at the fillet where high stress concentration exists together with highest tensile stress from bending or from the surface defects as shown in figure below. The crack slowly propagates over 80 to 90% of the life.

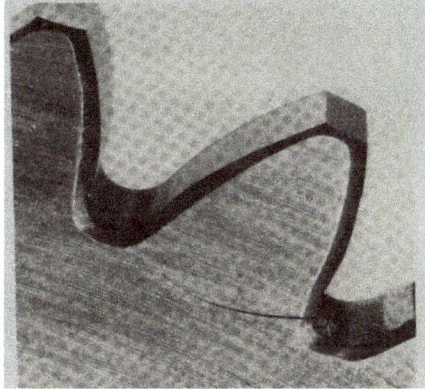

Root crack

Then crack propagates fast and suddenly results in fracture of the tooth as shown in figure below. The fractured surface will exhibit beach marks in the slow crack propagation region and brittle fracture behavior in sudden fracture region. Since time taken for the failure is very long, it is known as high cycle fatigue.

Tooth breakage

Tooth Breakage – High Cycle Fatigue

The tooth breakage in case of high cycle fatigue is shown in figure below.

High cycle fatigue

Tooth Breakage – Low Cycle Fatigue (Over Load)

Overload breakage or short (low) cycle fatigue causes stringy fibrous appearance in broken ductile material. In harder materials this break has a more silky or crystalline appearance as shown in figure below.

Low cycle fatigue (over load)

Tooth Breakage – Bending Fatigue

The figure below shows tooth fatigue by bending fatigue.

Bending fatigue

Tooth Breakage

Breakage is often due to load concentration along the tooth length as a result of errors in machining and assembly or of large elastic deformation of the shafts; tooth wear leading to weakening of the teeth results in increased dynamic loads. Shifting of sliding gears into mesh takes place without stopping the rotation of the shafts. Cracks are usually formed at the root of the teeth on the side of the stretched fibers where the highest tensile stresses occur together with local stresses due to the shape of the teeth. Fracture occurs mainly at a cross section through the root of the teeth.

In the case of fatigue failure, the fracture is of concave form in the body of the gear; it is of convex form when the failure is from overload. The teeth of herringbone or wide-face helical gears usually break off along a slanting cross section. To prevent tooth breakage, the beam strength of the gear teeth is checked by calculations. Fatigue pitting of the surface layers of the gear teeth is the most serious and widespread kind of tooth damage that may occur in gears even when they are enclosed, well lubricated and protected against dirt.

Gear Noise

The gear noise arises due to several reasons. At the contact point due to error in the gear profile, surface roughness, impact of tooth and sliding and rolling friction; bearings, churning of the lubricant, and windage.

The principal methods of combating noise are: improving the tooth finishing operations, changing over to helical gearing, modifying the profile by flanking, increasing the contact ratio, equalizing the load along the face width of the tooth rim, using crowned gears, and improving the design of the covers and housings.

Causes of Industrial Fires and Explosions

Here are five of the most common causes of industrial fires and explosions.

Combustible Dust

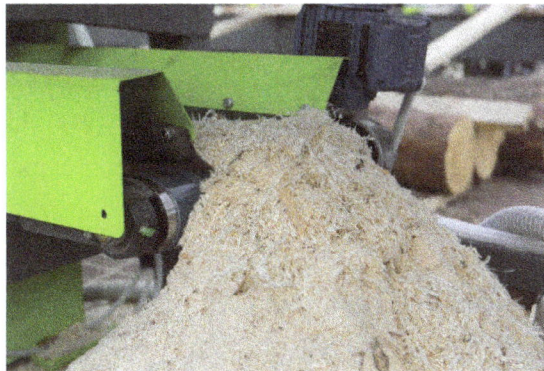

Often overlooked, and highly deadly, combustible dust is a major cause of fire in food manufacturing, woodworking, chemical manufacturing, metalworking, pharmaceuticals, and just about every other industry you can name. The reason is that just about everything, including food, dyes, chemicals, and metals — even materials that aren't fire risks in larger pieces — has the potential to be combustible in dust form.

And these explosions aren't easy to contain. In the typical incident, a small fire will result from combustible material coming into contact with an ignition source. This may be a dust explosion, but it doesn't have to be. In fact, it could be most any other type of explosion on this list.

However, this small explosion isn't the problem. The problem is what happens next. If there's dust in the area, the primary explosion will cause that dust to become airborne. Then, the dust cloud itself can ignite, causing a secondary explosion that can be many times the size and severity of the primary explosion. If enough dust has accumulated, these secondary explosions have the potential to bring down entire facilities, causing immense damage and fatalities.

Hot Work

Hot work is one of the leading causes of industrial fires across all industries.

Although hot work is commonly equated with welding and torch cutting, there are many other activities — including brazing, burning, heating, and soldering — that pose a fire hazard. This is because the sparks and molten material, which reach temperatures greater than 1000°F, can easily travel more than 35 feet.

Here are a few recent catastrophes that were the result of hot work:

- In 2014, a pier fire in California did more than $100 million in damage when it caused a partial collapse of a warehouse floor.

- In 2012, three workers performing hot work died disassembling a metal crude oil tank. The sparks from the work ignited vapors inside the tank, causing a fire that then spread to nearby woods.

- In 2010, one worker died and one was injured in an explosion while performing welding on a 10,000 gallon slurry tank. Similar to the previous incident, the sparks from the welding ignited vapors inside the tank.

Hot work is also a major culprit in combustible dust fires, as the sparks generated from the work can ignite dust in the surrounding area.

Prevention of Hot Work Incidents

Like combustible dust incidents, hot work disasters are preventable by following proper safety procedures.

- Avoid hot work if possible. This isn't always a feasible solution, but if there's an alternative, takes it.

- Train personnel on the hazards associated with hot work, any site-specific hazards, the proper policies and procedures, and the use of safety equipment.

- Ensure that the area is clear of flammable or combustible materials including dusts, liquids, and gases.

- Use a written permit system for all hot work projects, even where permits aren't required. Better safe than sorry.

- Supervise the work. Especially if you use outside contractors, make sure a safety professional is on hand to provide supervision.

Flammable Liquids and Gases

These fires, which often occur at chemical plants, can be disastrous.

Prevention of Flammable liquid and Gas Incidents

There is certainly some danger inherent in any work involving flammable liquids and gases, but all available safety precautions should be taken to mitigate these risks.

- Know the hazards. One major component of prevention is simply knowing the safety information for every liquid on your premises. This information is available on the material safety data sheet (MSDS) that comes with such products.

- Store flammable liquids properly. Make sure all hazardous materials are stored according to OSHA-compliant procedures.

- Control all ignition sources. Except for when you're intentionally heating the flammable materials, keep ignition sources as far away from them as possible.

- Provide personal protective equipment. This is a must across all categories of fire hazards but especially when liquids and gases are involved.

Equipment and Machinery

Faulty equipment and machinery are also major causes of industrial fires.

Heating and hot work equipment is typically the biggest problems here — in particular, furnaces that aren't properly installed, operated, and maintained. In addition, any mechanical equipment can become a fire hazard because of friction between the moving parts. This risk can be brought down to practically zero simply by following recommended cleaning and maintenance procedures, including lubrication.

What may surprise you is that even seemingly innocuous equipment can be a hazard under the right circumstances. And, in many cases, the equipment least likely to be thought of as a fire risk turns out to be the biggest problem. This is because companies may not recognize the risk and therefore won't take the necessary precautions.

Prevention of Equipment and Machinery Incidents

Strategies for preventing fires due to equipment and machinery issues fall into three main categories:

- Awareness
- Cleaning and housekeeping
- Maintenance

Awareness

You can't prevent risks you don't know exist. Neither can your employees. Provide safety awareness training so everyone in your facility knows what risks to watch out for and what to do if they find one.

Cleaning and housekeeping

Keep your equipment and machinery — and the area surrounding it — clean. Equipment, especially electrical equipment that is covered with dirt or grease constitutes a huge risk. By keeping your equipment and machinery clean, you'll up your chances that, should a fire start, it won't have enough fuel available to burn for long.

Maintenance

Finally, follow the manufacturer's recommended maintenance procedures for all of the equipment and machinery in your plant. In addition to reducing your fire risk by preventing overheating, regular maintenance will also keep your equipment working in tip-top shape.

Electrical hazards

Electrical fires are one of the top five causes of fires in manufacturing plants. Here a non-exhaustive list of specific electrical hazards:

- Wiring that is exposed or not up to code

- Overloaded outlets

- Extension cords

- Overloaded circuits

- Static discharge

The damage caused by these fires can quickly compound thanks to several of the other items on this list. Any of the above hazards can cause a spark, which can serve as an ignition source for combustible dust, as well as flammable liquids and gasses.

Prevention of Electrical Fire Incidents

As with the previous risks, the key to preventing electrical fires is awareness and prevention. This involves training, maintenance, and following best practices. Here are a few to put into practice right now:

- Don't overload electrical equipment or circuits.

- Don't leave temporary equipment plugged in when it's not in use.

- Avoid using extension cords, and never consider them permanent solutions.

- Use antistatic equipment where required by NFPA or OSHA.

- Follow a regular housekeeping plan to remove combustible dust and other hazardous materials from areas that contain equipment and machinery.

- Implement a reporting system so that anyone who observes an electrical fire risk can report it without consequences.

Failure Mode, Effects, and Criticality Analysis

Failure Mode, Effects & Criticality Analysis (FMECA) is a bottom-up (Hardware) or top-down.

Risk assessment it is inductive, or data-driven, linking elements of a failure chain as follows: Effect of Failure, Failure Mode and Causes/Mechanisms. These elements closely resemble the modern 5 Why technique in Root Cause Analysis (RCA). The Effect of Failure duplicates the experience of a user/customer and is then translated into the technical failure description or Failure Mode. The technical failure description answers the next question "Why", introducing causes that result in the failure mode. Each failure mode has a probability assigned and each cause has a failure rate assigned. If data is not available, probability of occurrence is assigned. The probability depends on the failure data source documents utilized in the FMECA. Unlike 5 Why, the FMECA is performed prior to any failure actually occurring. FMECA analyzes risk, which is measured by criticality (the combination of severity and probability), to take action and thus provide an opportunity to reduce the possibility of failure.

FMECA and Failure Mode and Effects Analysis (FMEA) are closely related tools. Each tool resolves to identify failure modes which may potentially cause product or process failure. FMEA is qualitative, exploring "what-if scenarios", where FMECA includes a degree of quantitative input taken from a source of known failure rates. A source for such data is Military Handbook 217 or equivalent.

There are two activities to perform FMECA:

1. Create the FMEA

2. Perform the Criticality Analysis

Measured criticality is the intersection of severity and cause probability rankings. Results are depicted in four primary criticality zones. Criticality is used to determine product or process design weaknesses. Two quantitative and one qualitative options exist for FMECA Criticality as identified below:

1. Quantitative

    •   Mode Criticality = Item Unreliability x Mode Ratio of Unreliability x Probability of Loss x Time (life).

    •   Item Criticality = Sum of Mode Criticalities.

2. Qualitative

    •   Compare failure modes via a Criticality Matrix, which identifies severity on the horizontal axis and qualitatively derived occurrence on the vertical axis.

    •   Note: Quality-One suggests a qualitative criticality matrix for the Quality-One Three Path Model for FMEA Development. Severity is on the vertical axis and occurrence is depicted on the horizontal axis. This is often used as an alternative for the Risk Priority Number (RPN) in FMEA.

Need to Perform Failure Mode, Effects & Criticality Analysis (FMECA)

The intent of the Failure Mode, Effects & Criticality Analysis methodology is to increase knowledge of risk and prevent failure. The tangible benefits of FMECA are offered in the following categories:

Design and Development Benefits

•   Increased reliability

•   Better quality

•   Higher safety margins

- Decreased development time and re-design

Operations Benefits

- More effective Control Plans

- Improved Verification and Validation testing requirements

- Optimized preventive and predictive maintenance

- Reliability growth analysis during product development

- Decreased waste and non-value added operations (Lean Operation and Manufacturing)

Cost Benefits

- Recognize failure modes in advance (when they are less costly to address)

- Minimized warranty costs

- Increased sales from customer satisfaction

Procedure of Failure Mode, Effects & Criticality Analysis (FMECA)

The basic assumption when performing FMECA instead of FMEA is the desire to have a more quantitative risk determination. The FMEA utilizes a more multi-functional team using guidelines to set Severity and Occurrence. The FMECA is performed by first completing an FMEA process worksheet and then completing the FMECA Criticality Worksheet.

The general steps for FMECA development are as follows:

- FMEA Portion

  - Define the system

  - Define ground rules and assumptions to help drive the design

  - Construct system Boundary Diagrams and Parameter Diagrams

  - Identify failure modes

  - Analyze failure effects

  - Determine causes of the failure modes

  - Feed results back into design process

- FMECA Portion

  - Transfer Information from the FMEA to the FMECA

  - Classify the failure effects by severity (change to FMECA severity)

  - Perform criticality calculations

o Rank failure mode criticality and determine highest risk items

o Take mitigation actions and document the remaining risk with rationale

o Follow-up on corrective action implementation/effectiveness

FMECA can often become time consuming and therefore available resources and team interest can be an issue as the process continues. Quality-One has developed the FMECA process below to utilize engineering resources effectively and ensure the FMECA has been developed thoroughly. The Quality-One approach is as follows:

Step 1: Perform the FMEA

The FMEA is a good starting place for the FMECA. FMEA allows for qualitative, and therefore creative, inputs from a multi-disciplined engineering team. FMEA provides the first inputs into design change and can jump start the risk mitigation process. The FMEA information is transferred into the FMECA Criticality Worksheet. The transferred data from the FMEA worksheet will include:

- Item Identification Number

- Item / Function

- Detailed Function and / or Requirements

- Failure Modes and Causes with Mechanisms of Failure

- Mission Phase or Operational Mode (DoD specific), often related to the Effects of Failure

Step 2: Determine Severity Level

Next, assign the Severity Level of each Effect of Failure. There are various severity tables to select from. The following is used in medical and some aerospace activities. The actual descriptions can be altered to fit any product or process design. There are generally four severity level classifications as follows:

- Catastrophic: Could result in death, permanent total disability, loss exceeding $1M, or irreversible severe environmental damage that violates law or regulation.

- Major/High Impact: Permanent partial disability, injuries or occupational illness resulting in hospitalization of 3 or more personnel, loss exceeding $200K but less than $1M, or reversible environmental damage causing a violation of law or regulation.

- Minor Impact: Could result in injury or occupational illness resulting in one or more lost work day(s), loss exceeding $10K but less than $200K, or mitigatable environmental damage without violation of law or regulation where restoration activities can be accomplished

- Low Impact: Result in minor injury or illness not resulting in a lost work day, loss exceeding $2K but less than $10K, or minimal environmental damage.

Step 3: Failure Effect Probability

In some applications of FMECA, a Beta value is assigned to the Failure Effect Probability. The

FMECA analyst may also use engineering judgement to determine the Beta value. The Beta / Effect Probability is placed in the FMECA Criticality Worksheet where:

- Actual Loss / 1.00

- Probable loss / >0.10 to <1.00

- Possible loss / >0 to =0.10

- No Effect / 0

A failure mode ratio is developed by assigning a proportion of the failure mode to each cause. The accumulation of all cause values equals 1.00.

Step 4: Probability of Occurrence (Quantitative)

Assign probability values for each Failure Mode, referencing the data source selected. Failure Probability and Failure Rate data can be found from several sources:

- Handbook 217 is referenced but any source of failure rate data can be used

- RAC databases, Concordia, etc.

If the Failure Mode probability is listed (functional approach) several columns of the FMECA Criticality Worksheet may be skipped. Criticality (Cr) can be calculated directly. When failure rates for failure modes and contributing components are desired, detailed failure rates for each component are assigned.

Next, we must assign Component Failure Rate (lambda). Failure Rates for each component are selected from the failure rate source document. Where there is no failure rate available, the qualitative values from the FMEA are used. FMEA may also be an alternative method on new or innovative designs.

Operating Time (t) represents the time or cycles the item or component will be expected to live. This is related to the expected duty cycle requirements.

Step 5: Calculate and Plot Criticality

In FMECA, Criticality is calculated in two ways:

- The Modal Criticality (each failure mode all causes) = Cm

- The Criticality of the Item (all failure modes summarized) = Cr

Formulas of each are not provided in this explanation but the essence of the elements of the calculation is as follows:

- Cm = The product of the following:
    - Failure Rate of the Part (lambda)
    - Failure Rate of the Effect (Beta)
    - Failure Mode Ratio (alpha)

o   Operating Time (units of time or cycles)

- Cr = The summation of all the Cm

Step 6: Design Feedback and Risk Mitigation

Risk mitigation is a discipline required to reduce possible failure. The identified risk in the criticality matrix is the substitute for failure and must be treated in the same context as a test failure or customer returned component or item. FMECA requires a change in risk levels / criticality after mitigation. A defect / defective detection strategy, commensurate to the risk level, may be required. Acceptable risk management strategy includes the following:

- Mitigation actions directed at Highest Severity and Probability combinations.

- Any risk where mitigation was unsuccessful is a candidate for Mistake Proofing or Quality Control, protecting the customer / consumer from the potential failure

    o   Detection methods are chosen for failure modes first and if possible individual causes which do not permit shipping or acceptance.

- Action logs and "risk registers" with revision history are kept for follow-up and closure of each undesirable risk.

Other examples of FMECA mitigation strategies to consider:

- Design change. Take a new direction on design technology, change components and/or review duty cycles for derating.

- Selection of a component with a lower lambda (failure rate). This can be expensive unless identified early in Product Development.

- Physical redundancy of the component. This option places the redundant component in a parallel configuration. Both must fail simultaneously for the failure mode to occur. If a safety concern exists, this option may require non-identical components.

- Software redundancy. The addition of a sensing circuit which can change the state of the product. This option often reduces the severity of the event by protecting components through duty cycle changes and reducing input stresses.

- Warning system. A placard and / or buzzer / light. This requires action by an operator or analyst to avoid a failure or the effect of failure.

- Detection and removal of the potential failure through testing or inspection. The inspection effectiveness must match the level of severity and criticality.

Step 7: Perform Maintainability Analysis

Maintainability Analysis looks at the highest risk items and determines which components will fail earliest. The cost and parts availability are also considered. This analysis can affect the location of the components or items when in the design phase. Design consideration must be given for quick access when serviceability is required more frequently.

- Access panels, easy to remove, permit service of the identified components and items. This can limit down time of important machinery.

- A spare parts list is typically created from the maintainability analysis.

Failure Rate

Failure rates are an important consideration in engineering. They are used to determine the reliability of a system or a component in a system. To calculate a failure rate, you need to observe the system or the component and record the time it takes to break down. As with any statistic, the more data you have, the more accurate the failure rate calculation will be. For example, if you were calculating the failure rate of a specific type of USB cable, your calculation would be more accurate if you tested 1,000 cables over a year rather than one cable over a few days.

Calculating Constant Failure Rates

In order to measure failure rates, you need a sample of identical components or systems that can be observed over time. For example, suppose you had five light bulbs connected to an automatic circuit that you could then turn on and off once per hour for 1,000 hours, giving you the following data:

- Bulb 1 burned out after 422 hours.

- Bulb 2 burned out after 744 hours

- Bulb 3 burned out after 803 hours

- Bulb 4 burned out after 678 hours

- Bulb 5 stayed lit for 1000 hours

This gives you 4 failures over a total of 3,647 hours.

To calculate the failure rate, divide the number of failures by the total number of hours, such as 4/3,647 = 0.0011 failures per hour.

In this example, the failure rate per hour is so small that it is almost insignificant. Multiplying the number by 1,000 would make it more meaningful to someone thinking about buying a light bulb, which would be 1.1 failures per 1,000 hours. Since there are 8,760 hours in one year, you can divide 3,647 by 8,760 to get 0.41 failures per year, or about 2 failures every five years.

Calculating MTBF

Another way to express failure rates is by using the Mean Time Between Failures. MTBF is usually used in high-quality systems where failures are expected to be rare and need to be minimized, like the guidance system on a commercial aircraft or the air bags in a passenger car. Knowing the MTBF allows manufacturers to recommend how often components should be inspected, maintained and replaced.

To calculate the MTBF, you divide the number of hours by the number of failures. In the case of the five light bulbs that were tested, which had a failure rate of 4 per 3,647, you determine the MTF as 3,647/4 = 909. The MTBF is therefore 909 hours.

Degrading Systems over Time

In most real-world scenarios, the likelihood of failure increases over time as components break down and parts wear out. A car's brake system, for example, is less likely to fail in the first year of ownership than it is after five years without maintenance. As a result, it is usually necessary for engineers to test components for longer periods of time and to calculate the failure rates for different intervals.

Fault Reporting

Fault Reporting is a maintenance concept to increase the operational availability and reduce operating cost.

Fault Report Life Cycle

The life cycle of a fault report involves three stages. Note that the term stage here refers to the life cycle of a fault rather than to the life cycle of a methodology.

- Test: The process begins in the Test Stage when a tester detects a defect and submits a fault report. Testing includes, for example, unit, integration, system and acceptance testing.

- Review: Upon submission, a reviewer (or a review team) examines the fault report in the Review Stage and either assigns it to a developer for correction, returns it to the Test Stage for more information, or rejects the fault report for a variety of reasons (e.g., works as intended, requires a change request, needs a decision from user).

- Correction: In the Correction Stage, the developer either corrects and subsequently returns the fault report to the Test Stage for retesting, returns the fault report to the Review Stage for reexamination or returns it to the Test Stage for further information.

Typical Life Cycle

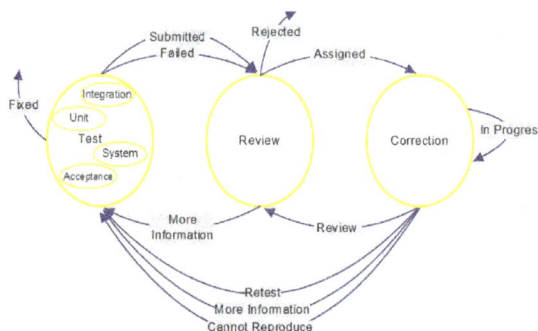

COMPLETE LIFE CYCLE OF A FAULT REPORT

The typical life cycle begins when a tester detects a defect and submits a fault report for review. A reviewer then reviews the fault report and assigns it to a developer for correction. The fault report remains assigned to the developer while correction is in progress. When correction is complete, the fault report is returned to Test for retest. The tester retests to verify that the defect has been fixed.

Retest Failure

When retesting shows that the defect has not been corrected, the fault report is resubmitted as a retest failure.

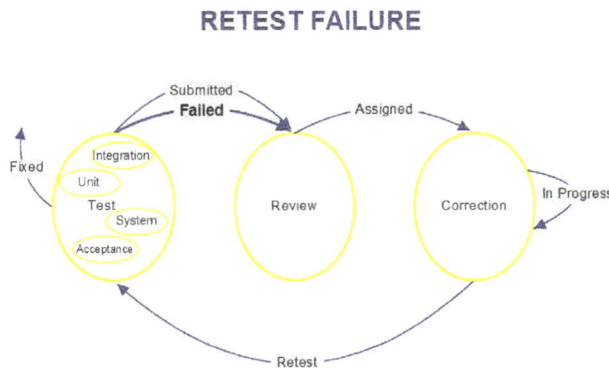

RETEST FAILURE

Rejection

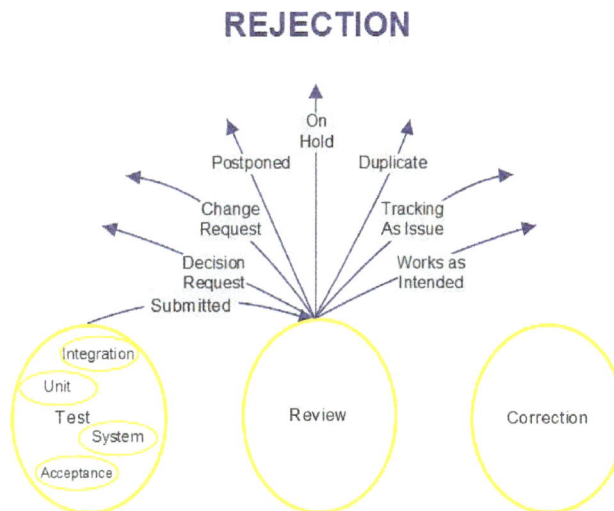

REJECTION

During review, a fault report may be rejected for various reasons including:

- requires a Decision Request to obtain clarification from the user community,

- requires a Change Request when the functionality described is beyond the scope of the requirements,

- requires postponement of correction when the functionality described is planned for implementation later in the project,

- requires placement on hold when the functionality described is not ready for testing (e.g., it is in unit test but a system test fault report is submitted),

- duplicates an existing fault report,

- describes an issue to be tracked rather than a defect,

- functionality works as intended and is according to design.

Reexamination

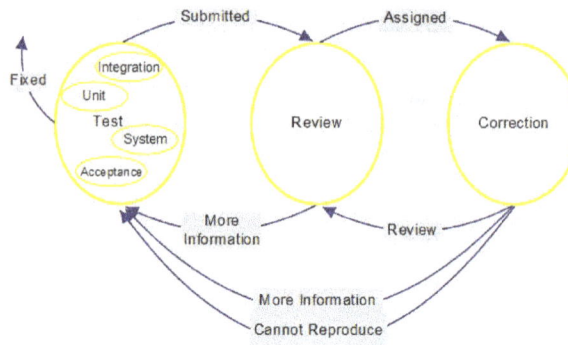

The fault report enters the reexamination cycle in the following scenarios. When the description of a fault report is inadequate for the purposes of review or correction, the report is returned to the tester for more information. In some cases, the developer assigned fault report may disagree with the reviewers conclusions and be returned for further review. When a fault cannot be reproduced by the assigned developer, it is returned to the tester for retesting. If the tester reproduces the defect, the fault report is re-submitted for review; otherwise, the defect is considered fixed.

References

- The-main-causes-of-industrial-machinery-failure: ybcomponents.co.uk, Retrieved 27 March 2018

- Failure-analysis-487: corrosionpedia.com, Retrieved 27 March 2018

- 5-major-causes-of-industrial-fires-explosions: news.nilfiskcfm.com, Retrieved 14 July 2018

- Calculate-failure-rates-6403358: sciencing.com, Retrieved 27 April 2018

- Defining-a-fault-report-life-cycle-111807: it.toolbox.com, Retrieved 20 June 2018

3

Lubrication

A lubricant is used to reduce friction and wear between two surfaces in contact. It can be a solid, liquid or gas. Adequate lubrication is essential for the smooth and continuous operation of machines and for prevention of excess stresses at bearings. This chapter has been carefully written to provide a comprehensive understanding of the stresses and strains that the different parts of a machine are subject to owing to friction and the lubrication that can be applied to reduce such effects. It analyzes the fundamental concepts of grease fitting, friction and wear, etc.

Friction

Friction is the force that resists the sliding or rolling of one solid object over another. Frictional forces, such as the traction needed to walk without slipping, may be beneficial, but they also present a great measure of opposition to motion. About 20 percent of the engine power of automobiles is consumed in overcoming frictional forces in the moving parts.

The major cause of friction between metals appears to be the forces of attraction, known as adhesion, between the contact regions of the surfaces, which are always microscopically irregular. Friction arises from shearing these "welded" junctions and from the action of the irregularities of the harder surface plowing across the softer surface.

Two simple experimental facts characterize the friction of sliding solids. First, the amount of friction is nearly independent of the area of contact. If a brick is pulled along a table, the frictional force is the same whether the brick is lying flat or standing on end. Second, friction is proportional to the load or weight that presses the surfaces together. If a pile of three bricks is pulled along a table, the friction is three times greater than if one brick is pulled. Thus, the ratio of friction F to load L is constant. This constant ratio is called the coefficient of friction and is usually symbolized by the Greek letter mu (μ). Mathematically, $\mu = F/L$. Because both friction and load are measured in units of force (such as pounds or newtons), the coefficient of friction is dimensionless. The value of the coefficient of friction for a case of one or more bricks sliding on a clean wooden table is about 0.5, which implies that a force equal to half the weight of the bricks is required just to overcome friction in keeping the bricks moving along at a constant speed. The frictional force itself is directed oppositely to the motion of the object. Because the friction thus far described arises between surfaces in relative motion, it is called kinetic friction.

Static friction, in contrast, acts between surfaces at rest with respect to each other. The value of static friction varies between zero and the smallest force needed to start motion. This smallest force required to start motion, or to overcome static friction, is always greater than the force required to continue the motion, or to overcome kinetic friction.

Rolling friction occurs when a wheel, ball, or cylinder rolls freely over a surface, as in ball and roller bearings. The main source of friction in rolling appears to be dissipation of energy involved in deformation of the objects. If a hard ball is rolling on a level surface, the ball is somewhat flattened and the level surface somewhat indented in the regions in contact. The elastic deformation or compression produced at the leading section of the area in contact is a hindrance to motion that is not fully compensated as the substances spring back to normal shape at the trailing section. The internal losses in the two substances are similar to those that keep a ball from bouncing back to the level from which it is dropped. Coefficients of sliding friction are generally 100 to 1,000 times greater than coefficients of rolling friction for corresponding materials. This advantage was realized historically with the transition from sledge to wheel.

Types of Friction

Static friction acts on objects when they are resting on a surface. For example, if you are hiking in the woods, there is static friction between your shoes and the trail each time you put down your foot. Without this static friction, your feet would slip out from under you, making it difficult to walk. In fact, that's exactly what happens if you try to walk on ice. That's because ice is very slippery and offers very little friction.

Sliding Friction

Sliding friction is friction that acts on objects when they are sliding over a surface. Sliding friction is weaker than static friction. That's why it's easier to slide a piece of furniture over the floor after you start it moving than it is to get it moving in the first place. Sliding friction can be useful. For example, you use sliding friction when you write with a pencil. The pencil "lead" slides easily over the paper, but there's just enough friction between the pencil and paper to leave a mark.

Rolling Friction

Rolling friction is friction that acts on objects when they are rolling over a surface. Rolling friction is much weaker than sliding friction or static friction. This explains why most forms of ground transportation use wheels, including bicycles, cars, 4-wheelers, roller skates, scooters, and skateboards. Ball bearings are another use of rolling friction. You can see what they look like in the Figure below. They let parts of a wheel or other machine roll rather than slide over on another.

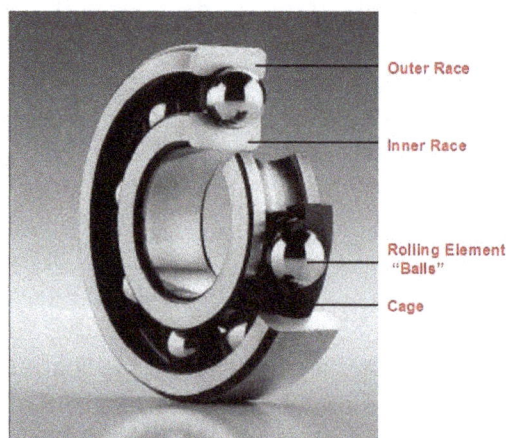

Outer Race

Inner Race

Rolling Element
"Balls"

Cage

Fluid Friction

Fluid friction is friction that acts on objects that are moving through a fluid. A fluid is a substance that can flow and take the shape of its container. Fluids include liquids and gases.

- Lubricated friction is a case of fluid friction where a lubricant fluid separates two solid surfaces.

- Skin friction is a component of drag, the force resisting the motion of a fluid across the surface of a body.

- Internal friction is the force resisting motion between the elements making up a solid material while it undergoes deformation.

When surfaces in contact move relative to each other, the friction between the two surfaces converts kinetic energy into thermal energy (that is, it converts work to heat). This property can have dramatic consequences, as illustrated by the use of friction created by rubbing pieces of wood together to start a fire. Kinetic energy is converted to thermal energy whenever motion with friction occurs, for example when a viscous fluid is stirred. Another important consequence of many types of friction can be wear, which may lead to performance degradation or damage to components. Friction is a component of the science of tribology.

Friction is desirable and important in supplying traction to facilitate motion on land. Most land vehicles rely on friction for acceleration, deceleration and changing direction. Sudden reductions in traction can cause loss of control and accidents.

Friction is not itself a fundamental force. Dry friction arises from a combination of inter-surface adhesion, surface roughness, surface deformation, and surface contamination. The complexity of these interactions makes the calculation of friction from first principles impractical and necessitates the use of empirical methods for analysis and the development of theory.

Friction is a non-conservative force - work done against friction is path dependent. In the presence of friction, some energy is always lost in the form of heat. Thus mechanical energy is not conserved.

Laws of Dry Friction

The elementary property of sliding (kinetic) friction were discovered by experiment in the 15th to 18th centuries and were expressed as three empirical laws:

- Amontons' First Law: The force of friction is directly proportional to the applied load.

- Amontons' Second Law: The force of friction is independent of the apparent area of contact.

- Coulomb's Law of Friction: Kinetic friction is independent of the sliding velocity.

Dry Friction

Dry friction resists relative lateral motion of two solid surfaces in contact. The two regimes of dry friction are 'static friction' ("stiction") between non-moving surfaces, and kinetic friction (sometimes called sliding friction or dynamic friction) between moving surfaces.

Coulomb friction, named after Charles-Augustin de Coulomb, is an approximate model used to calculate the force of dry friction. It is governed by the model:

$$F_f \leq \mu F_n$$

Where

- F_f is the force of friction exerted by each surface on the other. It is parallel to the surface, in a direction opposite to the net applied force.

- μ is the coefficient of friction, which is an empirical property of the contacting materials,

- F_n Is the normal force exerted by each surface on the other, directed perpendicular (normal) to the surface.

The Coulomb friction F_f may take any value from zero up to μF_n and the direction of the frictional force against a surface is opposite to the motion that surface would experience in the absence of friction. Thus, in the static case, the frictional force is exactly what it must be in order to prevent motion between the surfaces; it balances the net force tending to cause such motion. In this case, rather than providing an estimate of the actual frictional force, the Coulomb approximation provides a threshold value for this force, above which motion would commence. This maximum force is known as traction.

The force of friction is always exerted in a direction that opposes movement (for kinetic friction) or potential movement (for static friction) between the two surfaces. For example, a curling stone sliding along the ice experiences a kinetic force slowing it down. For an example of potential movement, the drive wheels of an accelerating car experience a frictional force pointing forward; if they did not, the wheels would spin, and the rubber would slide backwards along the pavement. Note that it is not the direction of movement of the vehicle they oppose, it is the direction of (potential) sliding between tire and road.

Normal Force

The normal force is defined as the net force compressing two parallel surfaces together; and its direction is perpendicular to the surfaces. In the simple case of a mass resting on a horizontal surface, the only component of the normal force is the force due to gravity, where $N = mg$. In this case, the magnitude of the friction force is the product of the mass of the object, the acceleration due to gravity, and the coefficient of friction. However, the coefficient of friction is not a function of mass or volume; it depends only on the material. For instance, a large aluminum block has the same coefficient of friction as a small aluminum block. However, the magnitude of the friction force itself depends on the normal force, and hence on the mass of the block.

If an object is on a level surface and the force tending to cause it to slide is horizontal, the normal force N between the object and the surface is just its weight, which is equal to its mass multiplied by the acceleration due to earth's gravity, g. If the object is on a tilted surface such as an inclined plane, the normal force is less, because less of the force of gravity is perpendicular to the face of the plane. Therefore, the normal force, and ultimately the frictional force, is determined using vector analysis, usually via a free body diagram. Depending on the situation, the calculation of the normal force may include forces other than gravity.

A block on a ramp

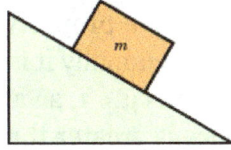

Free body diagram
of just the block

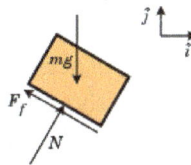

Free-body diagram for a block on a ramp. Arrows are vectors indicating directions and magnitudes of forces. N is the normal force, mg is the force of gravity, and F_f is the force of friction.

Coefficient of Friction

The coefficient of friction (COF), often symbolized by the Greek letter μ, is a dimensionless scalar value which describes the ratio of the force of friction between two bodies and the force pressing them together. The coefficient of friction depends on the materials used; for example, ice on steel has a low coefficient of friction, while rubber on pavement has a high coefficient of friction. Coefficients of friction range from near zero to greater than one. It is an axiom of the nature of friction between metal surfaces that it is greater between two surfaces of similar metals than between two surfaces of different metals— hence, brass will have a higher coefficient of friction when moved against brass, but less if moved against steel or aluminum.

For surfaces at rest relative to each other $\mu = \mu_s$ where μ_s is the coefficient of static friction. This is usually larger than its kinetic counterpart. The coefficient of static friction exhibited by a pair of contacting surfaces depends upon the combined effects of material deformation characteristics and surface roughness, both of which have their origins in the chemical bonding between atoms in each of the bulk materials and between the material surfaces and any adsorbed material. The fractality of surfaces, a parameter describing the scaling behavior of surface asperities, is known to play an important role in determining the magnitude of the static friction.

For surfaces in relative motion $\mu = \mu_k$, where μ_k is the coefficient of kinetic friction. The Coulomb friction is equal to F_f, and the frictional force on each surface is exerted in the direction opposite to its motion relative to the other surface.

Arthur Morin introduced the term and demonstrated the utility of the coefficient of friction. The coefficient of friction is an empirical measurement – it has to be measured experimentally, and cannot be found through calculations. Rougher surfaces tend to have higher effective values. Both static and kinetic coefficients of friction depend on the pair of surfaces in contact; for a given pair of surfaces, the coefficient of static friction is usually larger than that of kinetic friction; in some sets the two coefficients are equal, such as teflon-on-teflon.

Most dry materials in combination have friction coefficient values between 0.3 and 0.6. Values outside this range are rarer, but teflon, for example, can have a coefficient as low as 0.04. A value of zero would mean no friction at all, an elusive property. Rubber in contact with other surfaces can yield friction coefficients from 1 to 2. Occasionally it is maintained that μ is always < 1, but this is not true. While in most relevant applications μ < 1, a value above 1 merely implies that the force required to slide an object along the surface is greater than the normal force of the surface on the object. For example, silicone rubber or acrylic rubber-coated surfaces have a coefficient of friction that can be substantially larger than 1.

While it is often stated that the COF is a "material property," it is better categorized as a "system property." Unlike true material properties (such as conductivity, dielectric constant, yield strength), the COF for any two materials depends on system variables like temperature, velocity, atmosphere and also what are now popularly described as aging and deaging times; as well as on geometric properties of the interface between the materials, namely surface structure. For example, a copper pin sliding against a thick copper plate can have a COF that varies from 0.6 at low speeds (metal sliding against metal) to below 0.2 at high speeds when the copper surface begins to melt due to frictional heating. The latter speed, of course, does not determine the COF uniquely; if the pin diameter is increased so that the frictional heating is removed rapidly, the temperature drops, the pin remains solid and the COF rises to that of a 'low speed' test.

Approximate Coefficients of Friction

Materials		Static Friction, μ_s		Kinetic/Sliding Friction, μ_k	
Dry and clean		Lubricated	Dry and clean	Lubricated	
Aluminium	Steel	0.61		0.47	
Aluminum	Aluminum			1.5	
Gold	Gold			2.5	
Platinum	Platinum			3.0	
Silver	Silver			1.5	
Alumina ceramic	Silicon Nitride ceramic				0.004 (wet)
BAM (Ceramic alloy AlMgB$_{14}$)	Titanium boride (TiB$_2$)	0.04–0.05	0.02		
Brass	Steel	0.35-0.51	0.19	0.44	
Cast iron	Copper	1.05		0.29	
Cast iron	Zinc	0.85		0.21	
Concrete	Rubber	1.0	0.30 (wet)	0.6-0.85	0.45-0.75 (wet)
Concrete	Wood	0.62			
Copper	Glass	0.68			
Copper	Steel	0.53		0.36	
Glass	Glass	0.9-1.0		0.4	

Human synovial fluid	Cartilage		0.01		0.003
Ice	Ice	0.02-0.09			
Polyethene	Steel	0.2	0.2		
PTFE (Teflon)	PTFE (Teflon)	0.04	0.04		0.04
Steel	Ice	0.03			
Steel	PTFE (Teflon)	0.04-0.2	0.04		0.04
Steel	Steel	0.74-0.80	0.16	0.42-0.62	
Wood	Metal	0.2–0.6	0.2 (wet)		
Wood	Wood	0.25–0.5	0.2 (wet)		

Under certain conditions some materials have very low friction coefficients. An example is (highly ordered pyrolytic) graphite which can have a friction coefficient below 0.01. This ultralow-friction regime is called superlubricity.

Static Friction

Static friction is friction between two or more solid objects that are not moving relative to each other. For example, static friction can prevent an object from sliding down a sloped surface. The coefficient of static friction, typically denoted as μs, is usually higher than the coefficient of kinetic friction. Static friction is considered to arise as the result of surface roughness features across multiple length-scales at solid surfaces. These features, known as asperities are present down to nano-scale dimensions and result in true solid to solid contact existing only at a limited number of points accounting for only a fraction of the apparent or nominal contact area. The linearity between applied load and true contact area, arising from asperity deformation, gives rise to the linearity between static frictional force and normal force, found for typical Amonton-Coulomb type friction.

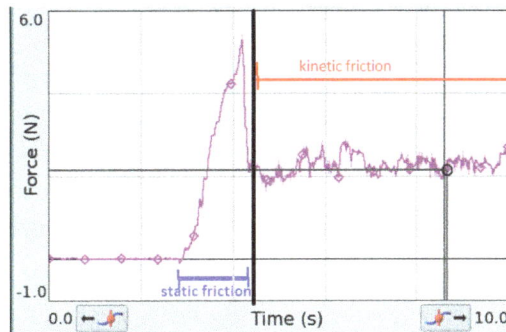

When the mass is not moving, the object experiences static friction. The friction increases as the applied force increases until the block moves. After the block moves, it experiences kinetic friction, which is less than the maximum static friction.

The static friction force must be overcome by an applied force before an object can move. The maximum possible friction force between two surfaces before sliding begins is the product of the coefficient of static friction and the normal force: $F_{max} = \mu_s F_n$. When there is no sliding occurring, the friction force can have any value from zero up to F_{max} ,. Any force smaller than F_{max} , attempting to slide one surface over the other is opposed by a frictional force of equal magnitude and opposite direction. Any force larger than F_{max} , overcomes the force of static friction and causes sliding to occur. The instant sliding occurs, static friction is no longer applicable—the friction between the two surfaces is then called kinetic friction.

An example of static friction is the force that prevents a car wheel from slipping as it rolls on the ground. Even though the wheel is in motion, the patch of the tire in contact with the ground is stationary relative to the ground, so it is static rather than kinetic friction.

The maximum value of static friction, when motion is impending, is sometimes referred to as limiting friction, although this term is not used universally.

An example of static friction is the force that prevents a car wheel from slipping as it rolls on the ground. Even though the wheel is in motion, the patch of the tire in contact with the ground is stationary relative to the ground, so it is static rather than kinetic friction.

The maximum value of static friction, when motion is impending, is sometimes referred to as limiting friction, although this term is not used universally.

Kinetic Friction

Kinetic friction, also known as dynamic friction or sliding friction, occurs when two objects are moving relative to each other and rub together (like a sled on the ground). The coefficient of kinetic friction is typically denoted as μk, and is usually less than the coefficient of static friction for the same materials. However, Richard Feynman comments that "with dry metals it is very hard to show any difference." The friction force between two surfaces after sliding begins is the product of the coefficient of kinetic friction and the normal force: $F_k = \mu_k F_n$

New models are beginning to show how kinetic friction can be greater than static friction. Kinetic friction is now understood, in many cases, to be primarily caused by chemical bonding between the surfaces, rather than interlocking asperities; however, in many other cases roughness effects are dominant, for example in rubber to road friction. Surface roughness and contact area affect kinetic friction for micro- and nano-scale objects where surface area forces dominate inertial forces.

The origin of kinetic friction at nanoscale can be explained by thermodynamics. Upon sliding, new surface forms at the back of a sliding true contact, and existing surface disappears at the front of it. Since all surfaces involve the thermodynamic surface energy, work must be spent in creating the new surface, and energy is released as heat in removing the surface. Thus, a force is required to move the back of the contact, and frictional heat is released at the front.

Angle of Friction

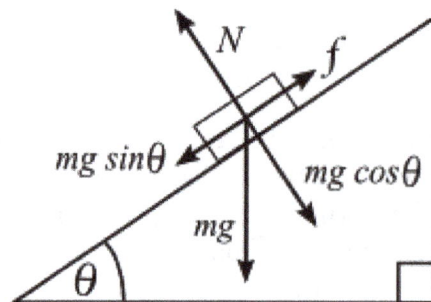

Angle of friction, θ, when block just starts to slide.

For certain applications it is more useful to define static friction in terms of the maximum angle before which one of the items will begin sliding. This is called the angle of friction or friction angle. It is defined as:

$$\tan\theta = \mu_s$$

Where θ is the angle from horizontal and μs is the static coefficient of friction between the objects. This formula can also be used to calculate μs from empirical measurements of the friction angle.

Friction at The Atomic Level

Determining the forces required to move atoms past each other is a challenge in designing nanomachines. In 2008 scientists for the first time were able to move a single atom across a surface, and measure the forces required. Using ultrahigh vacuum and nearly zero temperature (5 K), a modified atomic force microscope was used to drag a cobalt atom, and a carbon monoxide molecule, across surfaces of copper and platinum.

Limitations of the Coulomb Model

The Coulomb approximation mathematically follows from the assumptions that surfaces are in atomically close contact only over a small fraction of their overall area, that this contact area is proportional to the normal force (until saturation, which takes place when all area is in atomic contact), and that the frictional force is proportional to the applied normal force, independently of the contact area. Such reasoning aside, however, the approximation is fundamentally an empirical construct. It is a rule of thumb describing the approximate outcome of an extremely complicated physical interaction. The strength of the approximation is its simplicity and versatility. Though in general the relationship between normal force and frictional force is not exactly linear (and so the frictional force is not entirely independent of the contact area of the surfaces), the Coulomb approximation is an adequate representation of friction for the analysis of many physical systems.

When the surfaces are conjoined, Coulomb friction becomes a very poor approximation (for example, adhesive tape resists sliding even when there is no normal force, or a negative normal force). In this case, the frictional force may depend strongly on the area of contact. Some drag racing tires are adhesive for this reason. However, despite the complexity of the fundamental physics behind friction, the relationships are accurate enough to be useful in many applications.

"Negative" Coefficient of Friction

As of 2012, a single study has demonstrated the potential for an effectively negative coefficient of friction in the low-load regime, meaning that a decrease in normal force leads to an increase in friction. This contradicts everyday experience in which an increase in normal force leads to an increase in friction. This was reported in the journal Nature in October 2012 and involved the friction encountered by an atomic force microscope stylus when dragged across a graphene sheet in the presence of graphene-adsorbed oxygen.

Numerical Simulation of the Coulomb Model

Despite being a simplified model of friction, the Coulomb model is useful in many numerical simulation applications such as multibody systems and granular material. Even its most simple expression encapsulates the fundamental effects of sticking and sliding which are required in many applied cases, although specific algorithms have to be designed in order to efficiently numerically integrate mechanical systems with Coulomb friction and bilateral or unilateral contact. Some quite nonlinear effects, such as the so-called Painlevé paradoxes, may be encountered with Coulomb friction.

Dry friction and Instabilities

Dry friction can induce several types of instabilities in mechanical systems which display a stable behavior in the absence of friction. These instabilities may be caused by the decrease of the friction force with an increasing velocity of sliding, by material expansion due to heat generation during friction (the thermo-elastic instabilities), or by pure dynamic effects of sliding of two elastic materials (the Adams-Martins instabilities). The latter were originally discovered in 1995 by George G. Adams and João Arménio Correia Martins for smooth surfaces and were later found in periodic rough surfaces. In particular, friction-related dynamical instabilities are thought to be responsible for brake squeal and the 'song' of a glass harp, phenomena which involve stick and slip, modelled as a drop of friction coefficient with velocity.

A practically important case is the self-oscillation of the strings of bowed instruments such as the violin, cello, hurdy-gurdy, erhu, etc.

Frictional instabilities can lead to the formation of new self-organized patterns (or "secondary structures") at the sliding interface, such as in-situ formed tribofilms which are utilized for the reduction of friction and wear in so-called self-lubricating materials.

Fluid Friction

Fluid friction occurs between fluid layers that are moving relative to each other. This internal resistance to flow is named viscosity. In everyday terms, the viscosity of a fluid is described as its "thickness". Thus, water is "thin", having a lower viscosity, while honey is "thick", having a higher viscosity. The less viscous the fluid, the greater its ease of deformation or movement.

All real fluids (except superfluids) offer some resistance to shearing and therefore are viscous. For teaching and explanatory purposes it is helpful to use the concept of an inviscid fluid or an ideal fluid which offers no resistance to shearing and so is not viscous.

Lubricated Friction

Lubricated friction is a case of fluid friction where a fluid separates two solid surfaces. Lubrication is a technique employed to reduce wear of one or both surfaces in close proximity moving relative to each another by interposing a substance called a lubricant between the surfaces.

In most cases the applied load is carried by pressure generated within the fluid due to the frictional viscous resistance to motion of the lubricating fluid between the surfaces. Adequate lubrication allows smooth continuous operation of equipment, with only mild wear, and without excessive stresses or seizures at bearings. When lubrication breaks down, metal or other components can rub destructively over each other, causing heat and possibly damage or failure.

Skin Friction

Skin friction arises from the interaction between the fluid and the skin of the body, and is directly related to the area of the surface of the body that is in contact with the fluid. Skin friction follows the drag equation and rises with the square of the velocity.

Skin friction is caused by viscous drag in the boundary layer around the object. There are two ways to decrease skin friction: the first is to shape the moving body so that smooth flow is possible, like an airfoil. The second method is to decrease the length and cross-section of the moving object as much as is practicable.

Internal Friction

Internal friction is the force resisting motion between the elements making up a solid material while it undergoes deformation.

Plastic deformation in solids is an irreversible change in the internal molecular structure of an object. This change may be due to either (or both) an applied force or a change in temperature. The change of an object's shape is called strain. The force causing it is called stress.

Elastic deformation in solids is reversible change in the internal molecular structure of an object. Stress does not necessarily cause permanent change. As deformation occurs, internal forces oppose the applied force. If the applied stress is not too large these opposing forces may completely resist the applied force, allowing the object to assume a new equilibrium state and to return to its original shape when the force is removed. This is known as elastic deformation or elasticity.

Radiation Friction

As a consequence of light pressure, Einstein in 1909 predicted the existence of "radiation friction" which would oppose the movement of matter. He wrote, "radiation will exert pressure on both sides of the plate. The forces of pressure exerted on the two sides are equal if the plate is at rest. However, if it is in motion, more radiation will be reflected on the surface that is ahead during the motion (front surface) than on the back surface. The backwardacting force of pressure exerted on the front surface is thus larger than the force of pressure acting on the back. Hence, as the resultant of the two forces, there remains a force that counteracts the motion of the plate and that increases with the velocity of the plate. We will call this resultant 'radiation friction' in brief".

Other Types of friction

Braking Friction

Any wheel equipped with a brake is capable of generating a large retarding force, usually for the purpose of slowing and stopping a vehicle or piece of rotating machinery. Braking friction differs from rolling friction because the coefficient of friction for rolling friction is small whereas the coefficient of friction for braking friction is designed to be large by choice of materials for brake pads.

Triboelectric Effect

Rubbing dissimilar materials against one another can cause a build-up of electrostatic charge, which can be hazardous if flammable gases or vapors are present. When the static build-up discharges, explosions can be caused by ignition of the flammable mixture.

Belt Friction

Belt friction is a physical property observed from the forces acting on a belt wrapped around a pulley, when one end is being pulled. The resulting tension, which acts on both ends of the belt, can be modeled by the belt friction equation.

In practice, the theoretical tension acting on the belt or rope calculated by the belt friction equation can be compared to the maximum tension the belt can support. This helps a designer of such a rig to know how many times the belt or rope must be wrapped around the pulley to prevent it from slipping. Mountain climbers and sailing crews demonstrate a standard knowledge of belt friction when accomplishing basic tasks.

Reducing Friction

Devices

Devices such as wheels, ball bearings, roller bearings, and air cushion or other types of fluid bearings can change sliding friction into a much smaller type of rolling friction.

Many thermoplastic materials such as nylon, HDPE and PTFE are commonly used in low friction bearings. They are especially useful because the coefficient of friction falls with increasing imposed load. For improved wear resistance, very high molecular weight grades are usually specified for heavy duty or critical bearings.

Lubricants

A common way to reduce friction is by using a lubricant, such as oil, water, or grease, which is placed between the two surfaces, often dramatically lessening the coefficient of friction. The science of friction and lubrication is called tribology. Lubricant technology is when lubricants are mixed with the application of science, especially to industrial or commercial objectives.

Super lubricity, a recently discovered effect, has been observed in graphite: it is the substantial decrease of friction between two sliding objects, approaching zero levels. A very small amount of frictional energy would still be dissipated.

Lubricants to overcome friction need not always be thin, turbulent fluids or powdery solids such as graphite and talc; acoustic lubrication actually uses sound as a lubricant.

Another way to reduce friction between two parts is to superimpose micro-scale vibration to one of the parts. This can be sinusoidal vibration as used in ultrasound-assisted cutting or vibration noise, known as dither.

Energy of Friction

According to the law of conservation of energy, no energy is destroyed due to friction, though it may be lost to the system of concern. Energy is transformed from other forms into thermal energy. A sliding hockey puck comes to rest because friction converts its kinetic energy into heat which raises the thermal energy of the puck and the ice surface. Since heat quickly dissipates, many early philosophers, including Aristotle, wrongly concluded that moving objects lose energy without a driving force.

When an object is pushed along a surface along a path C, the energy converted to heat is given by a line integral, in accordance with the definition of work

$$E_{th} = \int_C \mathbf{F}_{fric}\, \mathbf{x} \cdot d\mathbf{x} = \int_C \mu_k \mathbf{F}_n\, \mathbf{x} \cdot d\mathbf{x}$$

Where

$\mathbf{F}_{fric}$ is the friction force,

$\mathbf{F}_n$ is the vector obtained by multiplying the magnitude of the normal force by a unit vector pointing against the object's motion,

μ_k is the coefficient of kinetic friction, which is inside the integral because it may vary from location to location (e.g. if the material changes along the path),

$\mathbf{x}$ is the position of the object.

Energy lost to a system as a result of friction is a classic example of thermodynamic irreversibility.

Work of Friction

In the reference frame of the interface between two surfaces, static friction does no work, because there is never displacement between the surfaces. In the same reference frame, kinetic friction is always in the direction opposite the motion, and does negative work. However, friction can do positive work in certain frames of reference. One can see this by placing a heavy box on a rug, then pulling on the rug quickly. In this case, the box slides backwards relative to the rug, but moves forward relative to the frame of reference in which the floor is stationary. Thus, the kinetic friction between the box and rug accelerates the box in the same direction that the box moves, doing positive work.

The work done by friction can translate into deformation, wear, and heat that can affect the contact surface properties (even the coefficient of friction between the surfaces). This can be beneficial as in polishing. The work of friction is used to mix and join materials such as in the process of friction welding. Excessive erosion or wear of mating sliding surfaces occurs when work due to frictional forces rise to unacceptable levels. Harder corrosion particles caught between mating surfaces in relative motion (fretting) exacerbates wear of frictional forces. Bearing seizure or failure may result from excessive wear due to work of friction. As surfaces are worn by work due to friction, fit and surface finish of an object may degrade until it no longer functions properly.

Applications

Friction is an important factor in many engineering disciplines.

Transportation

- Automobile brakes inherently rely on friction, slowing a vehicle by converting its kinetic energy into heat. Incidentally, dispersing this large amount of heat safely is one technical challenge in designing brake systems. Disk brakes rely on friction between a disc and brake pads that are squeezed transversely against the rotating disc. In Drum brakes, brake shoes or pads are pressed outwards against a rotating cylinder (brake drum) to create friction. Since braking discs can be more efficiently cooled than drums, disc brakes have better stopping performance.

- Rail adhesion refers to the grip wheels of a train have on the rails.

- Road slipperiness is an important design and safety factor for automobiles

  o   Split friction is a particularly dangerous condition arising due to varying friction on either side of a car.

  o   Road texture affects the interaction of tires and the driving surface.

Measurement

- A tribometer is an instrument that measures friction on a surface.

- A profilograph is a device used to measure pavement surface roughness.

Household usage

- Friction is used to heat and ignite matchsticks (friction between the head of a matchstick and the rubbing surface of the match box).

- Sticky pads are used to prevent object from slipping off smooth surfaces by effectively increasing the friction coefficient between the surface and the object.

Wear

Wear is a process of interaction between surfaces, which causes the deformation and removal of material on the surfaces due to the effect of mechanical action between the sliding faces. Wear also refers to the dimension loss of plastic deformation. Plastic deformation leads to wear; it causes the deterioration of metal surfaces, which is known as "metallic wear".

Wear is the result of many things such corrosion, erosion, abrasion, chemical processes, or combinations of these factors. The processes of wear are studied in the field of tribology.

Types of Wear

Three types of wear are abrasion, adhesion, and corrosion. Each type has inherent problems and benefits that can be affected by materials, lubrication, and surface finish.

Abrasive

There are two common types: two-body and three-body abrasion. Two-body abrasion refers to surfaces that slide across each other where the one (hard) material will dig in and remove some of the other (soft) material. An example of two-body abrasion is using a file to shape a work piece. Three-body abrasion is where particles between the two surfaces remove material from one or both surfaces. The tumbling process is an example of this.

Tumbling involves using particles to sand and polish the surface of a part. The particles that cause abrasion are often called contaminants. Contaminants are anything that enters a system that creates abrasion. While lubrication is imperative, an active lubrication system can introduce contaminants that cause abrasion. Filters remove contaminants and are one of the reasons proper maintenance and replacement of filters is important. However, the lubrication, or the additives in it, can react with the metal, creating a thin monolayer of contaminants that also make proper lubrication selection important to reduce wear to your equipment.

Abrasive wear can have benefits, such as water jetting. Water jetting has the ability to cut through metal with relative ease. This can reduce property changes that can occur with other processes that generate excessive heat while cutting.

Surface roughness is another important variable for wear. Two-body abrasion is reduced by having smoother surface roughness. For example, a journal or sleeve bearing made out of a softer material will slide against a harder drive shaft with little to no abrasion due to the surface finish. Using materials with similar hardness is generally not advised. The reason for the softer bearing material is to further reduce wear. Contaminants can become embedded into the softer materials and stop three-body abrasion from occurring. This technique might damage the bearing, but is preferred as it is designed to be relatively easy and more cost-effective to replace than a drive shaft. The rougher surfaces can increase the coefficient of friction and micro-peaks can break off, contributing to contaminants that are related to abrasion.

Adhesion

Surface roughness also contributes to adhesion. For this type of wear a material's compatibility will be important. Compatibility does not mean materials that work well together; rather, that the materials "like" each other, causing them to stick together. This compatibility forms a bond causing parts to seize and even become cold-welded together. There are a few general rules to follow for material selection to make sure unwanted adhesive wear doesn't occur. Materials that make contact with one another, in general, should:

- Not dissolve in the other;

- Not, in given environment and other conditions, form into an alloy;

- Not be identical (e.g., an aluminum shaft with an aluminum bearing);

- Have at least one metal from the B-subgroup (e.g., elements to the right of Nickel, Palladium, and Platinum on the periodic table).

Adhesion is possible to calculate. The adhesion and abrasive wear calculations share the same formula; however, it can vary by as much as +/-20%. This inaccuracy is due to constant changing surface conditions and lubrication during operation. It may be better than no data, but designers need to be aware of the limitations and accuracy of the formula. Trying to calculate or predict wear is made more difficult if components have non-conforming geometries, such as when gear teeth and cams are involved. These components can have difficulty staying properly lubricated. To reduce adhesive wear, sometimes corrosive wear is purposely induced.

Corrosive

Chlorides, phosphates, or sulfides can be added to induce corrosion and reduce a more destructive adhesive wear. Corrosive wear is more often thought of as something you want to prevent. Rust, or oxidation, is the No. 1 form of corrosive wear. Lubrication, material selection, surface finish, including coatings—like in abrasive and adhesive wear—are the main factors to consider.

Noble materials are noted as having non-corrosive properties. Gold specifically is used in electronics as a coating due to its ability to resist corrosion. Noble materials are often used sparingly or in minimal waste processes due to their cost. Other materials are self-anodizing. Aluminum is known for reacting with oxygen to form a layer of aluminum oxide that will prevent oxidation.

Iron and ferrous materials not only are prone to rust, but will flake off, exposing another layer allowing the oxidation to continue the degradation process. In the case of self-anodizing materials, slight abrasions or even stresses can cause crack propagation or scrap off the aluminum oxide, allowing the corrosive wear to continue.

Stress affects corrosive wear. Stress corrosion and corrosion fatigue will significantly accelerate corrosive wear. The difference between these types is the loading situation. Static loads cause stress corrosion where more dynamic loading, such as cyclic loads will cause corrosion fatigue.

Fretting Wear

Fretting wear is the repeated cyclical rubbing between two surfaces. Over a period of time fretting which will remove material from one or both surfaces in contact. It occurs typically in bearings, although most bearings have their surfaces hardened to resist the problem. Another problem occurs when cracks in either surface are created, known as fretting fatigue. It is the more serious of the two phenomena because it can lead to catastrophic failure of the bearing. An associated problem occurs when the small particles removed by wear are oxidised in air. The oxides are usually harder than the underlying metal, so wear accelerates as the harder particles abrade the metal surfaces further. Fretting corrosion acts in the same way, especially when water is present. Unprotected bearings on large structures like bridges can suffer serious degradation in behavior, especially when salt is used them during winter to deice the highways carried by the bridges. The problem of fretting corrosion was involved in the Silver Bridge tragedy and the Mianus River Bridge accident.

Erosive Wear

Erosive wear can be defined as an extremely short sliding motion and is executed within a short time interval. Erosive wear is caused by the impact of particles of solid or liquid against the surface of an object. The impacting particles gradually remove material from the surface through repeated deformations and cutting actions. It is a widely encountered mechanism in industry. Due to the nature of the conveying process, piping systems are prone to wear when abrasive particles have to be transported.

The rate of erosive wear is dependent upon a number of factors. The material characteristics of the particles, such as their shape, hardness, and impact velocity and impingement angle are primary factors along with the properties of the surface being eroded. The impingement angle is one of the most important factors and is widely recognized in literature. For ductile materials, the maximum wear rate is found when the impingement angle is approximately 30°, whilst for non-ductile materials the maximum wear rate occurs when the impingement angle is normal to the surface.

Corrosion and Oxidation Wear

Corrosion and oxidation wear occurs both in lubricated and unlubricated contacts. The fundamental cause are chemical reactions between the worn material and the corroding medium. Wear caused by a synergistic action of tribological stresses and corrosion is also called tribocorrosion.

Wear Stages

Under nominal operation conditions, the wear rate normally changes in three different stages:

- Primary stage or early run-in period, where surfaces adapt to each other and the wear-rate might vary between high and low.

- Secondary stage or mid-age process, where steady wear can be observed. Most of the component's operational life is spent in this stage.

- Tertiary stage or old-age period, where surfaces are subjected to rapid failure due to a high rate of wear.

Note that the wear rate is strongly influenced by the operating conditions and the formation of tribofilms. The secondary stage is shortened with increasing severity of environmental conditions, such as high temperatures, strain rates and stresses.

So-called wear maps, demonstrating wear rate under different operation condition, are used to determine stable operation points for tribological contacts. Wear maps also show dominating wear modes under different loading conditions.

In explicit wear tests simulating industrial conditions between metallic surfaces, there are no clear chronological distinction between different wear-stages due to big overlaps and symbiotic relations between various friction mechanisms. Surface engineering and treatments are used to minimize wear and extend the components working life.

Wear Testing

Several standard test methods exist for different types of wear to determine the amount of material removal during a specified time period under well-defined conditions. ASTM International Committee G-2 standardizes wear testing for specific applications, which are periodically updated. The Society for Tribology and Lubrication Engineers (STLE) has documented a large number of frictional, wear and lubrication tests. Standardized wear tests are used to create comparative material rankings for a specific set of test parameter as stipulated in the test description. To obtain more accurate predictions of wear in industrial applications it is necessary to conduct wear testing under conditions simulating the exact wear process.

An attrition test is a test that is carried out to measure the resistance of a granular material to wear.

Modeling of Wear

The Reye–Archard–Khrushchov wear law is the classic wear prediction model.

Measuring Wear

Wear Coefficient

The wear coefficient is a physical coefficient used to measure, characterize and correlate the wear of materials.

Mass Loss vs. Volume Loss

Volume loss is a more accurate measure of wear than mass loss, particularly when comparing the wear resistance properties of materials with large differences in density. For example, a weight loss of 14 g in a sample composed of tungsten carbide and cobalt (density = 14000 kg/m^3) and a weight loss of 2.7 g in a similar sample of aluminium alloy (density = 2700 kg/m^3) both result in the same level of wear (1 cm^3) when expressed as a volume loss.

Lubricant Analysis

Lubricant analysis is an alternative, indirect way of measuring wear. Here, wear is detected by the presence of wear particles in a liquid lubricant. To gain further insights into the nature of the particles, chemical (such as XRF, ICP-OES), structural (such as ferrography) or optical analysis (such as light microscopy) can be performed.

Lubrication

Lubrication is the introduction of any of various substances between sliding surfaces to reduce wear and friction. Nature has been applying lubrication since the evolution of synovial fluid, which lubricates the joints and bursas of vertebrate animals. Prehistoric people used mud and reeds to lubricate sledges for dragging game or timbers and rocks for construction. Animal fat lubricated the axles of the first wagons and continued in wide use until the petroleum industry arose in the 19th century, after which crude oil became the chief source of lubricants. The natural lubricating capacity of crude oil has been steadily improved through the development of a wide variety of products designed for the specific lubricating needs of the automobile, the airplane, the diesel locomotive, the turbojet, and power machinery of every description. The improvements in petroleum lubricants have in turn made possible the increase in speed and capacity of industrial and other machinery.

Lubricants containing oil have additives that enhance, add or suppress properties within the base oil. The amount of additives depends on the type of oil and the application for which it will be used. For instance, engine oil might have a dispersant added.

A dispersant keeps insoluble matter conglomerated together to be removed by the filter upon circulation. In environments that undergo extremes in temperature, from cold to hot, a viscosity index (VI) improver may be added. These additives are long organic molecules that stay bunched together in cold conditions and unravel in hotter environments.

This process changes the oil's viscosity and allows it to flow better in cold conditions while still maintaining its high-temperature properties. The only problem with additives is that they can be depleted, and in order to restore them back to sufficient levels, generally the oil volume must be replaced.

The Role of a Lubricant

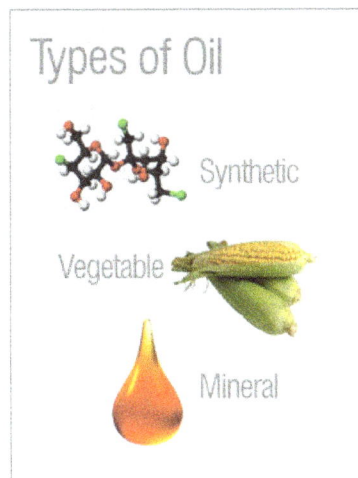

Types of Oil

Synthetic

Vegetable

Mineral

The primary functions of a lubricant are to:

- Reduce friction

- Prevent wear

- Protect the equipment from corrosion

- Control temperature (dissipate heat)

- Control contamination (carry contaminants to a filter or sump)

- Transmit power (hydraulics)

- Provide a fluid seal

Sometimes the functions of reducing friction and preventing wear are used interchangeably. However, friction is the resistance to motion, and wear is the loss of material as a result of friction, contact fatigue and corrosion. There is a significant difference. In fact, not all that causes friction (e.g., fluid friction) causes wear, and not all that causes wear (e.g., cavitational erosion) causes friction.

Reducing friction is a key objective of lubrication, but there are many other benefits of this process. Lubricating films can help prevent corrosion by protecting the surface from water and other corrosive substances. In addition, they play an important role in controlling contamination within systems.

The lubricant works as a conduit in which it transports contaminants to filters to be removed. These fluids also aid in temperature control by absorbing heat from surfaces and transferring it to a point of lower temperature where it can be dissipated.

Types of Lubrication

There are three basic varieties of lubrication: fluid-film, boundary, and solid.

Fluid-Film Lubrication.

Interposing a fluid film that completely separates sliding surfaces results in this type of lubrication. The fluid may be introduced intentionally, as the oil in the main bearings of an automobile, or unintentionally, as in the case of water between a smooth rubber tire and a wet pavement. Although the fluid is usually a liquid, it may also be a gas. The gas most commonly employed is air.

To keep the parts separated, it is necessary that the pressure within the lubricating film balance the load on the sliding surfaces. If the lubricating film's pressure is supplied by an external source, the system is said to be lubricated hydrostatically. If the pressure between the surfaces is generated as a result of the shape and motion of the surfaces themselves, however, the system is hydrodynamically lubricated. This second type of lubrication depends upon the viscous properties of the lubricant.

Boundary Lubrication

A condition that lies between unlubricated sliding and fluid-film lubrication is referred to as boundary lubrication, also defined as that condition of lubrication in which the friction between surfaces is determined by the properties of the surfaces and properties of the lubricant other than viscosity. Boundary lubrication encompasses a significant portion of lubrication phenomena and commonly occurs during the starting and stopping of machines.

Solid Lubrication

Solids such as graphite and molybdenum disulfide are widely used when normal lubricants do not possess sufficient resistance to load or temperature extremes. But lubricants need not take only such familiar forms as fats, powders, and gases; even some metals commonly serve as sliding surfaces in some sophisticated machines.

A lubricant primarily controls friction and wear, but it can and ordinarily does perform numerous other functions, which vary with the application and usually are interrelated.

Control Functions

The amount and character of the lubricant made available to sliding surfaces have a profound effect upon the friction that is encountered. For example, disregarding such related factors as heat and wear but considering friction alone between two oil-film lubricated surfaces, the friction can be 200 times less than that between the same surfaces with no lubricant. Under fluid-film conditions, friction is directly proportional to the viscosity of the fluid. Some lubricants, such as petroleum derivatives, are available in a great range of viscosities and thus can satisfy a broad spectrum of functional requirements. Under boundary lubrication conditions, the effect of viscosity on friction becomes less significant than the chemical nature of the lubricant. Delicate instruments, for example, must not be lubricated with fluids that would attack and corrode the finer metals.

lubricant	relative viscosity (air = 1)	typical minimum film thickness in bearing applications (in.)	typical unit load in bearing applications (lb per sq in.)
air	1	0.00005–0.0004	1–10
water	33	0.0004–0.001	25–75
oil	1,000	0.002–0.004	200–500
Characteristics of three typical lubricants			

Wear occurs on lubricated surfaces by abrasion, corrosion, and solid-to-solid contact. Proper lubricants will help combat each type. They reduce abrasive and solid-to-solid contact wear by providing a film that increases the distance between the sliding surfaces, thereby lessening the damage by abrasive contaminants and surface asperities. The role of a lubricant in controlling corrosion of surfaces is twofold. When machinery is idle, the lubricant acts as a preservative. When machinery is in use, the lubricant controls corrosion by coating lubricated parts with a protective film that may contain additives to neutralize corrosive materials. The ability of a lubricant to control corrosion is directly related to the thickness of the lubricant film remaining on the metal surfaces and the chemical composition of the lubricant.

Lubricants also can assist in controlling temperature by reducing friction and carrying off the heat that is generated. Effectiveness depends upon the amount of lubricant supplied, the ambient temperature, and the provision for external cooling. To a lesser extent, the type of lubricant also affects surface temperature.

Other Functions

Various lubricants are employed as hydraulic fluids in fluid transmission devices. Others can be used to remove contaminants in mechanical systems. Detergent-dispersant additives, for instance, suspend sludges and remove them from the sliding surfaces of internal-combustion engines.

In specialized applications such as transformers and switchgear, lubricants with high dielectric constants act as electrical insulators. For maximum insulating properties, a lubricant must be kept free of contaminants and water. Lubricants also act as shock-damping fluids in energy-transferring devices (e.g., shock absorbers) and around such machine parts as gears that are subjected to high intermittent loads.

Liquid, Oily Lubricants

Animal and vegetable products were certainly man's first lubricants and were used in large quantities. But, because they lack chemical inertness and because lubrication requirements have become more demanding, they have been largely superseded by petroleum products and by synthetic materials. Some organic substances such as lard oil and sperm oil are still in use as additives because of their special lubricating properties.

Petroleum lubricants are predominantly hydrocarbon products extracted from fluids that occur naturally within the Earth. They are used widely as lubricants because they possess a combination of the following desirable properties: (1) availability in suitable viscosities, (2) low volatility, (3) inertness (resistance to deterioration of the lubricant), (4) corrosion protection (resistance to deterioration of the sliding surfaces), and (5) low cost.

Synthetic lubricants generally can be characterized as oily, neutral liquid materials not usually obtained directly from petroleum but having some properties similar to petroleum lubricants. In certain ways they are superior to hydrocarbon products. Synthetics exhibit greater stability of viscosity with temperature changes, resistance to scuffing and oxidation, and fire resistance. Since the properties of synthetics vary considerably, each synthetic lubricant tends to find a special application.

synthetic lubricant	typical uses
dibasic acid esters	instrument oil, jet turbine lubricant, hydraulic fluid
phosphate esters	fire-resistant hydraulic fluid, low-temperature lubricant
silicones	damping fluid, low-volatility grease base
silicate esters	heat transfer fluid, high-temperature hydraulic fluid
polyglycol ether compounds	synthetic engine oil, hydraulic fluids, forming and drawing compounds
fluorol compounds	nonflammable fluid, extreme oxidation-resistant lubricant
Synthetic lubricants and typical applications	

Another form of oily lubricant is grease, a solid or semisolid substance consisting of a thickening agent in a liquid lubricant. Soaps of aluminum, barium, calcium, lithium, sodium, and strontium are the major thickening agents. Nonsoap thickeners consist of such inorganic compounds as modified clays or fine silicas, or such organic materials as arylureas or phthalocyanine pigments.

Lubrication by grease may prove more desirable than lubrication by oil under conditions when (1) less frequent lubricant application is necessary, (2) grease acts as a seal against loss of lubricant and ingress of contaminants, (3) less dripping or splattering of lubricant is called for, or (4) less sensitivity to inaccuracies in the mating parts is needed.

Solid Lubricants

A solid lubricant is a film of solid material composed of inorganic or organic compounds or of metal.

There are three general kinds of inorganic compounds that serve as solid lubricants:

1. Layer-lattice solids: materials such as graphite and molybdenum disulfide, commonly called molysulfide, have a crystal lattice structure arranged in layers. Strong bonds between atoms within a layer and relatively weak bonds between atoms of different layers allow the lamina to slide on one another. Other such materials are tungsten disulfide, mica, boron nitride, borax, silver sulfate, cadmium iodide, and lead iodide. Graphite's low friction is due largely to adsorbed films; in the absence of water vapor, graphite loses its lubricating properties and becomes abrasive. Both graphite and molysulfide are chemically inert and have high thermal stability.

2. Miscellaneous soft solids: a variety of inorganic solids such as white lead, lime, talc, bentonite, silver iodide, and lead monoxide are used as lubricants.

3. Chemical conversion coatings: many inorganic compounds can be formed on a metallic surface by chemical reaction. The best known such lubricating coatings are sulfide, chloride, oxide, phosphate, and oxalate films.

Solid organic lubricants are usually divided into two broad classes:

1. Soaps, waxes, and fats: this class includes metallic soaps of calcium, sodium, lithium; animal waxes (e.g., beeswax and spermaceti wax); fatty acids (e.g., stearic and palmitic acids); and fatty esters (e.g., lard and tallow).

2. Polymeric films: these are synthetic substances such as polytetrafluoroethylene and polychlorofluoroethylene. One major advantage of such film-type lubricants is their resistance to deterioration during exposure to the elements. Thus, 1/2-inch- (1.3-centimetre-) thick plates of polymeric film are used in modern prestressed concrete construction to permit thermal movement of beams resting atop columns. Such expansion and contraction of the structural members is facilitated by the long-lived polymeric film plate.

Thin films of soft metal on a hard substrate can act as effective lubricants, if the adhesion to the substrate is good. Such metals include lead, tin, and indium.

Gaseous Lubricants

Lubrication with a gas is analogous in many respects to lubrication with a liquid, since the same principles of fluid-film lubrication apply. Although both gases and liquids are viscous fluids, they differ in two important particulars. The viscosity of gases is much lower and the compressibility much greater than for liquids. Film thicknesses and load capacities therefore are much lower with a gas such as air. In equipment that handles gases of various kinds, it is often desirable to lubricate the sliding surfaces with gas in order to simplify the apparatus and reduce contamination to and from the lubricant. The list of gases used in this manner is extensive and includes air, steam, industrial gases, and liquid-metal vapors.

With so many types of materials capable of acting as lubricants under certain conditions, coverage of the properties of all of them is impractical. Mention is made only of those properties usually considered characteristic of commercially significant fluid lubricants.

Viscosity

Of all the properties of fluid lubricants, viscosity is the most important, since it determines the amount of friction that will be encountered between sliding surfaces and whether a thick enough film can be built up to avoid wear from solid-to-solid contact. Viscosity customarily is measured by a viscometer, which determines the flow rate of the lubricant under standard conditions; the higher the flow rate, the lower the viscosity. The rate is expressed in centipoises, reyns, or seconds Saybolt universal (SSU) depending, respectively, upon whether metric, English, or commercial units are used. In most liquids, viscosity drops appreciably as the temperature is raised. Since little change of viscosity with fluctuations in temperature is desirable to keep variations in friction at a minimum, fluids often are rated in terms of viscosity index. The less the viscosity is changed by temperature, the higher the viscosity index.

Pour Point

The pour point, or the temperature at which a lubricant ceases to flow, is important in appraising flow properties at low temperature. As such, it can become the determining factor in selecting one lubricant from among a group with otherwise identical properties.

Flash Point

The flash point, or the temperature at which a lubricant momentarily flashes in the pressure of a test flame, aids in evaluating fire-resistance properties. Like the pour-point factor, the flash point may in some instances become the major consideration in selecting the proper lubricant, especially in lubricating machinery handling highly flammable material.

Oiliness

Oiliness generally connotes relative ability to operate under boundary lubrication conditions. The term relates to a lubricant's tendency to wet and adhere to a surface. There is no formal test for the measurement of oiliness; determination of this factor is chiefly through subjective judgment and experience. The most desirable lubricant for a specific use need not necessarily be the oiliest; e.g., long-fibre grease, which is low in oiliness as compared with machine oils, is usually preferable for packing rolling bearings.

Neutralization Number

The neutralization number is a measure of the acid or alkaline content of new oils and an indicator of the degree of oxidation degradation of used oils. This value is ascertained by titration, a standard analytical chemical technique, and is defined as the number of milligrams of potassium hydroxide required to neutralize one gram of the lubricant.

Penetration Number

The penetration number, applied to grease, is a measure of the film characteristics of the grease. The test consists of dropping a standard cone into the sample of grease being tested. Gradations indicate the depth of penetration: the higher the number, the more fluid the grease.

Grease Fitting

Grease fittings are also known as zerk fittings, grease nipples, or Alemite fittings.

A Zerk fitting is a small metal fitting used in many mechanical systems for the purpose of feeding lubricants such as grease.

The grease is fed into the bearing or mechanical system under moderate or high-pressure with the use of a grease gun.

Zerk fittings are permanently installed into the bearing by a threaded connection. A nipple connection is left for the purpose of attaching a grease gun to feed lubricants into the bearing when required.

Working of Zerk Fittings

A Zerk fitting consists of a small captive bearing ball which moves back and forth against the force of a retaining spring. The pressure supplied by the grease gun will force this ball to move back.

This arrangement allows the lubricant to pass through a channel. When the pressure ceases, the captive ball will return to its closed position.

This ball helps prevent dirt from entering the system, and it functions similar to a check valve to prevent grease from escaping back to the fitting. The ball can be wiped clean to reduce the amount of debris carried into the bearing with the grease.

A Zerk fitting has a convex shape which allows the concave tip of the grease gun to seal against the fitting.

Tools for Greasing Zerk Fittings

- Standard Size Grease Guns: These tools hold a full tube of grease and offer enough leverage to make it easy to produce pressure when pushing grease into stubborn fittings. Unfortunately, they are unwieldy when crawling under things and almost require three hands to operate. These are great when they have a long hose and a swivel or 90° head.

- Mini Pistol Grip Guns: These small and agile grease guns are great for crawling around and under equipment, but they run out faster because they hold much less grease.

- Electric Grease Guns: This is the cat's meow when you grease Zerk fittings. Use a cordless grease gun when you're going to be greasing a lot of fittings, or when your hands just don't work like they used to work.

- Rejuvenator: Sometimes neglected Zerk fittings seize or plug up. There are tools meant to clear these fittings usually called "grease fitting tools" or "fitting rejuvenators." Typically they are two piece affairs that require you to load them with grease or diesel fuel, place them on the fitting and then strike them with a hammer to produce lots of pressure to clear the blockage. Sometimes they work, sometimes they don't. Good ones are not cheap, and cheap ones are not good, generally speaking. If the Zerk is in a bear of a spot, then a rejuvenator may be your best bet.

- Replacement Zerks: Your local auto parts store, tractor dealer or farm store will likely offer an assortment pack of Zerk grease fittings.

Tips for Greasing Zerk Fittings

- Listen for the Crack: When you grease Zerk fittings, listen for the crack. Once you have filled a void full of grease, the seals on either end usually make a crackling sound as they give way to let the overabundance of grease exit the joint. Stop before you blow out the seal.

- Use Just Enough: Don't overfill when you grease Zerk fittings. Usually three or four pumps of grease is enough and over-greasing a joint pushes grease out the aforementioned seals, which attracts dust, sand, and dirt. Contaminated grease can damage moving parts, so avoid having excess squish out the seals.

- Keep them clean: Carry a rag to clean off the Zerk after you grease. Again, exposed grease attracts dust, sand, and dirt. Avoid pushing contaminated grease into your fitting next time by cleaning it off when you grease.

- Choose the Right Product: Not all greases are created equal. Find which kind of grease is recommended by the manufacturer for that fitting. Does it need a low-temp or high-temp grease? Crude base or synthetic? When in doubt, check.

- Consider Compatibility: Not all greases are compatible. Don't mix greases because they don't all play well together. Be sure to stay consistent since mixing the wrong greases can cause reactions that result in more harm than good.

- Wear Gloves: Disposable exam or mechanic's gloves are perfect for greasing Zerk fittings, because you're bound to get your hands covered in grease.

References

- Beer, Ferdinand P.; Johnston, E. Russel, Jr. (1996). Vector Mechanics for Engineers(Sixth ed.). McGraw-Hill. p. 397. ISBN 0-07-297688-8

- Friction, science: britannica.com, Retrieved 09 July 2018

- Hutchings, Ian M. (2016-08-15). "Leonardo da Vinci's studies of friction". Wear. 360–361: 51–66. doi:10.1016/j.wear.2016.04.019

- Wear-abrasion-1169: corrosionpedia.com, Retrieved 20 May 2018

- Kirk, Tom (July 22, 2016). "Study reveals Leonardo da Vinci's 'irrelevant' scribbles mark the spot where he first recorded the laws of friction". phys.org. Retrieved 2016-07-26

- What-s-difference-between-types-wear, materials: machinedesign.com, Retrieved 24 April 2018

- Adams, G. G. (1995). "Self-excited oscillations of two elastic half-spaces sliding with a constant coefficient of friction". Journal of Applied Mechanics. 62: 867–872. Bibcode:1995JAM....62..867A. doi:10.1115/1.2896013

- Lubrication, technology: britannica.com, Retrieved 28 May 2018

- Meriam, J. L.; Kraige, L. G. (2002). Engineering Mechanics (fifth ed.). John Wiley & Sons. p. 328. ISBN 0-471-60293-0

- What-is-lubrication-28766: machinerylubrication.com, Retrieved 18 July 2018

- Popova, Elena; Popov, Valentin L. (2015-06-01). "The research works of Coulomb and Amontons and generalized laws of friction". Friction. 3 (2): 183–190. doi:10.1007/s40544-015-0074-6

- What-are-zerk-fittings-or-grease-fittings: bearingsinc.info, Retrieved 10 March 2018

- "Ultra-low friction coefficient in alumina–silicon nitride pair lubricated with water". Wear. 296: 656–659. doi:10.1016/j.wear.2012.07.030. Retrieved 2015-04-27

Bearings

A bearing is an element of a machine that either allows or prohibits motion of machine elements or reduces friction between moving parts. The topics elaborated in this chapter address the different types of bearings that are used in machinery, such as plain bearing, ball bearing, roller bearing, air bearing and spiral groove bearing.

A bearing is a device that is used to enable rotational or linear movement, while reducing friction and handling stress. Resembling wheels, bearings literally enable devices to roll, which reduces the friction between the surface of the bearing and the surface it's rolling over. It's significantly easier to move, both in a rotary or linear fashion, when friction is reduced—this also enhances speed and efficiency.

Bearings make use of a relatively simple structure: a ball with internal and external smooth metal surfaces, to aid in rolling. The ball itself carries the weight of the load—the force of the load's weight is what drives the bearing's rotation. However, not all loads put force on a bearing in the same manner. There are two different kinds of loading: radial and thrust.

A radial load, as in a pulley, simply puts weight on the bearing in a manner that causes the bearing to roll or rotate as a result of tension. A thrust load is significantly different, and puts stress on the bearing in an entirely different way. If a bearing (think of a tire) is flipped on its side (think now of a tire swing) and subject to complete force at that angle (think of three children sitting on the tire swing), this is called thrust load. A bearing that is used to support a bar stool is an example of a bearing that is subject only to thrust load.

Many bearings are prone to experiencing both radial and thrust loads. Car tires, for example, carry a radial load when driving in a straight line: the tires roll forward in a rotational manner as a result of tension and the weight they are supporting. However, when a car goes around a corner, it is subject to thrust load because the tires are no longer moving solely in a radial fashion and cornering force weighs on the side of the bearing.

Types of Bearings

There are two types of bearings, contact and noncontact. Contact-type bearings have mechanical contact between elements, and they include sliding, rolling, and flexural bearings. Mechanical contact means that stiffness normal to the direction of motion can be very high, but wear or fatigue can limit their life.

Non-contact bearings include externally pressurized and hydrodynamic fluid-film (liquid, air, mixed phase) and magnetic bearings. The lack of mechanical contact means that static friction can be eliminated, although viscous drag occurs when fluids are present; however, life can be virtually infinite if the external power units required to operate them do not fail.

Each type of bearing has its own niche application area, and thus design engineers must be familiar with different types of bearings, and their applications and limitations.

Ball Bearings

Ball bearings are extremely common because they can handle both radial and thrust loads, but can only handle a small amount of weight. They are found in a wide array of applications, such as roller blades and even hard drives, but are prone to deforming if they are overloaded.

Roller Bearings

Roller bearings are designed to carry heavy loads—the primary roller is a cylinder, which means the load is distributed over a larger area, enabling the bearing to handle larger amounts of weight. This structure, however, means the bearing can handle primarily radial loads, but is not suited to thrust loads. For applications where space is an issue, a needle bearing can be used. Needle bearings work with small diameter cylinders, so they are easier to fit in smaller applications.

Ball Thrust Bearings

These kinds of bearings are designed to handle almost exclusively thrust loads in low-speed low-weight applications. Bar stools, for example, make use of ball thrust bearings to support the seat.

Roller Thrust Bearings

Roller thrust bearings, much like ball thrust bearings, handle thrust loads. The difference, however, lies in the amount of weight the bearing can handle: roller thrust bearings can support significantly larger amounts of thrust load, and are therefore found in car transmissions, where they are used to support helical gears. Gear support in general is a common application for roller thrust bearings.

Tapered Roller Bearings

This style of bearing is designed to handle large radial and thrust loads—as a result of their load versatility, they are found in car hubs due to the extreme amount of both radial and thrust loads that car wheels are expected to carry.

Specialized Bearings

There are, of course, several kinds of bearings that are manufactured for specific applications, such as magnetic bearings and giant roller bearings. Magnetic bearings are found in high-speed devices because it has no moving parts—this stability enables it to support devices that move unconscionably fast. Giant roller bearings are used to move extremely large and heavy loads, such as buildings and large structural components.

Motions

Common motions permitted by bearings are:

- Radial rotation e.g. shaft rotation

- Linear motion e.g. drawer

- Spherical rotation e.g. ball and socket joint

- Hinge motion e.g. door, elbow, knee

Friction

Reducing friction in bearings is often important for efficiency, to reduce wear and to facilitate extended use at high speeds and to avoid overheating and premature failure of the bearing. Essentially, a bearing can reduce friction by virtue of its shape, by its material, or by introducing and containing a fluid between surfaces or by separating the surfaces with an electromagnetic field.

- By shape, gains advantage usually by using spheres or rollers, or by forming flexure bearings.

- By material, exploits the nature of the bearing material used. (An example would be using plastics that have low surface friction).

- By fluid, exploits the low viscosity of a layer of fluid, such as a lubricant or as a pressurized medium to keep the two solid parts from touching, or by reducing the normal force between them.

- By fields, exploits electromagnetic fields, such as magnetic fields, to keep solid parts from touching.

- Air pressure exploits air pressure to keep solid parts from touching.

Combinations of these can even be employed within the same bearing. An example of this is where the cage is made of plastic, and it separates the rollers/balls, which reduce friction by their shape and finish.

Loads

Bearing design varies depending on the size and directions of the forces that they are required to support. Forces can be predominately radial, axial (thrust bearings), or bending moments perpendicular to the main axis.

Speeds

Different bearing types have different operating speed limits. Speed is typically specified as maximum relative surface speeds, often specified ft/s or m/s. Rotational bearings typically describe performance in terms of the product DN where D is the mean diameter (often in mm) of the bearing and N is the rotation rate in revolutions per minute.

Generally there is considerable speed range overlap between bearing types. Plain bearings typically handle only lower speeds, rolling element bearings are faster, followed by fluid bearings and finally magnetic bearings which are limited ultimately by centripetal force overcoming material strength.

Play

Some applications apply bearing loads from varying directions and accept only limited play or "slop" as the applied load changes. One source of motion is gaps or "play" in the bearing. For example, a 10 mm shaft in a 12 mm hole has 2 mm play.

Allowable play varies greatly depending on the use. As example, a wheelbarrow wheel supports radial and axial loads. Axial loads may be hundreds of newtons force left or right, and it is typically acceptable for the wheel to wobble by as much as 10 mm under the varying load. In contrast, a lathe may position a cutting tool to ±0.002 mm using a ball lead screw held by rotating bearings. The bearings support axial loads of thousands of newtons in either direction, and must hold the ball lead screw to ±0.002 mm across that range of loads.

Stiffness

A second source of motion is elasticity in the bearing itself. For example, the balls in a ball bearing are like stiff rubber, and under load deform from round to a slightly flattened shape. The race is also elastic and develops a slight dent where the ball presses on it.

The stiffness of a bearing is how the distance between the parts which are separated by the bearing varies with applied load. With rolling element bearings this is due to the strain of the ball and race. With fluid bearings it is due to how the pressure of the fluid varies with the gap (when correctly loaded, fluid bearings are typically stiffer than rolling element bearings).

Service Life

Fluid and Magnetic Bearings

Fluid and magnetic bearings can have practically indefinite service lives. In practice, there are fluids bearings supporting high loads in hydroelectric plants that have been in nearly continuous service since about 1900 and which show no signs of wear.

Rolling Element Bearings

Rolling element bearing life is determined by load, temperature, maintenance, lubrication, material defects, contamination, handling, installation and other factors. These factors can all have a significant effect on bearing life. For example, the service life of bearings in one application was extended dramatically by changing how the bearings were stored before installation and use, as vibrations during storage caused lubricant failure even when the only load on the bearing was its own weight; the resulting damage is often false brinelling. Bearing life is statistical: several samples of a given bearing will often exhibit a bell curve of service life, with a few samples showing significantly better or worse life. Bearing life varies because microscopic structure and contamination vary greatly even where macroscopically they seem identical.

L10 Life

Bearings are often specified to give an "L10" life (outside the USA, it may be referred to as "B10" life.) This is the life at which ten percent of the bearings in that application can be expected to have failed due to classical fatigue failure (and not any other mode of failure like lubrication starvation, wrong mounting etc.), or, alternatively, the life at which ninety percent will still be operating. The L10 life of the bearing is theoretical life and may not represent service life of the bearing. Bearings are also rated using Co (static loading) value. This is the basic load rating as a reference, and not an actual load value.

Plain Bearings

For plain bearings, some materials give much longer life than others. Some of the John Harrison clocks still operate after hundreds of years because of the lignum vitae wood employed in their construction, whereas his metal clocks are seldom run due to potential wear.

Flexure Bearings

Flexure bearings rely on elastic properties of material. Flexure bearings bend a piece of material repeatedly. Some materials fail after repeated bending, even at low loads, but careful material selection and bearing design can make flexure bearing life indefinite.

Short-life Bearings

Although long bearing life is often desirable, it is sometimes not necessary. Tedric A. Harris describes a bearing for a rocket motor oxygen pump that gave several hours life, far in excess of the several tens of minutes life needed.

Composite Bearings

Depending on the customized specifications (backing material and PTFE compounds), composite bearings can operate up to 30 years without maintenance.

Oscillating Bearings

For bearings which are used in oscillating applications, customized approaches to calculate L10 are used.

External Factors

The service life of the bearing is affected by many parameters that are not controlled by the bearing manufacturers. For example, bearing mounting, temperature, exposure to external environment, lubricant cleanliness and electrical currents through bearings etc. High frequency PWM inverters can induce currents in a bearing, which can be suppressed by use of ferrite chokes.

The temperature and terrain of the micro-surface will determine the amount of friction by the touching of solid parts.

Certain elements and fields reduce friction, while increasing speeds.

Strength and mobility help determine the amount of load the bearing type can carry.

Alignment factors can play a damaging role in wear and tear, yet overcome by computer aid signaling and non-rubbing bearing types, such as magnetic levitation or air field pressure.

Maintenance and Lubrication

Many bearings require periodic maintenance to prevent premature failure, but many others require little maintenance. The latter include various kinds of fluid and magnetic bearings, as well as rolling-element bearings that are described with terms including sealed bearing and sealed for life. These contain seals to keep the dirt out and the grease in. They work successfully in many applications, providing maintenance-free operation. Some applications cannot use them effectively.

Nonsealed bearings often have a grease fitting, for periodic lubrication with a grease gun, or an oil cup for periodic filling with oil. Before the 1970s, sealed bearings were not encountered on most machinery, and oiling and greasing were a more common activity than they are today. For example, automotive chassis used to require "lube jobs" nearly as often as engine oil changes, but today's car chassis are mostly sealed for life. From the late 1700s through mid-1900s, industry relied on many workers called oilers to lubricate machinery frequently with oil cans.

Factory machines today usually have lube systems, in which a central pump serves periodic charges of oil or grease from a reservoir through lube lines to the various lube points in the machine's bearing surfaces, bearing journals, pillow blocks, and so on. The timing and number of such lube cycles is controlled by the machine's computerized control, such as PLC or CNC, as well as by manual override functions when occasionally needed. This automated process is how all modern CNC machine tools and many other modern factory machines are lubricated. Similar lube systems are also used on nonautomated machines, in which case there is a hand pump that a machine operator is supposed to pump once daily (for machines in constant use) or once weekly. These are called one-shot systems from their chief selling point: one pull on one handle to lube the whole machine, instead of a dozen pumps of an alemite gun or oil can in a dozen different positions around the machine.

The oiling system inside a modern automotive or truck engine is similar in concept to the lube systems mentioned above, except that oil is pumped continuously. Much of this oil flows through passages drilled or cast into the engine block and cylinder heads, escaping through ports directly onto bearings, and squirting elsewhere to provide an oil bath. The oil pump simply pumps constantly, and any excess pumped oil continuously escapes through a relief valve back into the sump.

Many bearings in high-cycle industrial operations need periodic lubrication and cleaning, and many require occasional adjustment, such as pre-load adjustment, to minimize the effects of wear.

Bearing life is often much better when the bearing is kept clean and well lubricated. However, many applications make good maintenance difficult. For example, bearings in the conveyor of a rock crusher are exposed continually to hard abrasive particles. Cleaning is of little use, because cleaning is expensive yet the bearing is contaminated again as soon as the conveyor resumes operation. Thus, a good maintenance program might lubricate the bearings frequently but not include any disassembly for cleaning. The frequent lubrication, by its nature, provides a limited kind of cleaning action, by displacing older (grit-filled) oil or grease with a fresh charge, which itself collects grit before being displaced by the next cycle.

Rolling-element Bearing outer Race Fault Detection

Rolling-element bearings are widely used in the industries today, and hence maintenance of these bearings becomes an important task for the maintenance professionals. The rolling-element bearings wear out easily due to metal-to-metal contact, which creates faults in the outer race, inner race and ball. It is also the most vulnerable component of a machine because it is often under high load and high running speed conditions. Regular diagnostics of rolling-element bearing faults is critical for industrial safety and operations of the machines along with reducing the maintenance costs or avoiding shutdown time. Among the outer race, inner race and ball, the outer race tends to be more vulnerable to faults and defects.

There is still room for discussion as to whether the rolling element excites the natural frequencies of bearing component when it passes the fault on the outer race. Hence we need to identify the bearing outer race natural frequency and its harmonics. The bearing faults create impulses and results in strong harmonics of the fault frequencies in the spectrum of vibration signals. These fault frequencies are sometimes masked by adjacent frequencies in the spectra due to their little energy. Hence, a very high spectral resolution is often needed to identify these frequencies during a FFT analysis. The natural frequencies of a rolling element bearing with the free boundary conditions are 3 kHz. Therefore, in order to use the bearing component resonance bandwidth method to detect the bearing fault at an initial stage a high frequency range accelerometer should be adopted, and data obtained from a long duration needs to be acquired. A fault characteristic frequency can only be identified when the fault extent is severe, such as that of a presence of a hole in the outer race. The harmonics of fault frequency is a more sensitive indicator of a bearing outer race fault. For a more serious detection of defected bearing faults waveform, spectrum and envelope techniques will help reveal these faults. However, if a high frequency demodulation is used in the envelope analysis in order to detect bearing fault characteristic frequencies, the maintenance professionals have to be more careful in the analysis because of resonance, as it may or may not contain fault frequency components.

Using spectral analysis as a tool to identify the faults in the bearings faces challenges due to issues like low energy, signal smearing, cyclostationarity etc. High resolution is often desired to differentiate the fault frequency components from the other high-amplitude adjacent frequencies. Hence, when the signal is sampled for FFT analysis, the sample length should be large enough to give adequate frequency resolution in the spectrum. Also, keeping the computation time and memory within limits and avoiding unwanted aliasing may be demanding. However, a minimal frequency resolution required can be obtained by estimating the bearing fault frequencies and other vibration frequency components and its harmonics due to shaft speed, misalignment, line frequency, gearbox etc.

Packing

Some bearings use a thick grease for lubrication, which is pushed into the gaps between the bearing surfaces, also known as packing. The grease is held in place by a plastic, leather, or rubber gasket (also called a gland) that covers the inside and outside edges of the bearing race to keep the grease from escaping.

Bearings may also be packed with other materials. Historically, the wheels on railroad cars used sleeve bearings packed with waste or loose scraps of cotton or wool fiber soaked in oil, and then later used solid pads of cotton.

Ring oiler

Bearings can be lubricated by a metal ring that rides loosely on the central rotating shaft of the bearing. The ring hangs down into a chamber containing lubricating oil. As the bearing rotates, viscous adhesion draws oil up the ring and onto the shaft, where the oil migrates into the bearing to lubricate it. Excess oil is flung off and collects in the pool again.

Splash Lubrication

Some machines contain a pool of lubricant in the bottom, with gears partially immersed in the liquid, or crank rods that can swing down into the pool as the device operates. The spinning wheels fling oil into the air around them, while the crank rods slap at the surface of the oil, splashing it randomly on the interior surfaces of the engine. Some small internal combustion engines specifically contain special plastic flinger wheels which randomly scatter oil around the interior of the mechanism.

Pressure Lubrication

For high speed and high power machines, a loss of lubricant can result in rapid bearing heating and damage due to friction. Also in dirty environments the oil can become contaminated with dust or debris that increases friction. In these applications, a fresh supply of lubricant can be continuously supplied to the bearing and all other contact surfaces, and the excess can be collected for filtration, cooling, and possibly reuse. Pressure oiling is commonly used in large and complex internal combustion engines in parts of the engine where directly splashed oil cannot reach, such as up into overhead valve assemblies. High speed turbochargers also typically require a pressurized oil system to cool the bearings and keep them from burning up due to the heat from the turbine.

Composite bearings

Composite bearings are designed with a self-lubricating polytetrafluroethylene (PTFE) liner with a laminated metal backing. The PTFE liner offers consistent, controlled friction as well as durability whilst the metal backing ensures the composite bearing is robust and capable of withstanding high loads and stresses throughout its long life. Its design also makes it lightweight-one tenth the weight of a traditional rolling element bearing.

Plain Bearing

Plain bearings are cylindrical sleeves that bear light to moderate radial loads. They slide radially or axially over shafts to allow rotary motion or linear motion (or sometimes both) of these loads. Plain bearings of all types are compact and lightweight with high strength-to-weight ratio.

Plain bearings have none of the moving parts that rolling-element bearings have, so minimize fail points; they're also cost effective for even fairly rugged applications. Common variations are metallic sleeve bearings (which often ride loads on a hydrodynamic or full film of lubrication) and self-lubricating plastic bearings in an array of geometries for bushing, thrust bearing, and integral-slide applications.

Plain-bearing ratings are based in part on test results and its material modulus of elasticity, flexural strength, shore-D hardness, maximum surface pressure and running speed, rotating, and maximum load capacity — with the latter related to the plain bearing's material compressive limit. (Here, recall that the compressive limit is the point at which 0.2% permanent deformation occurs).

In addition, a pressure-speed (PV) value expresses plain-bearing load capacity — usually in in psi times the shaft rpm. However, note that PV values are only one to help determine a plain bearing's overall load capacity — especially where a PV expression might mislead engineers into thinking that a plain bearing can bear excessively high loads if the speed is very low. In other words, use of PV values requires concurrent consideration of real-world speed and load limits.

Types of Plane Bearings

Sleeve bearings to Sleeve bearings are the most-common type of plane bearing, and support linear, oscillating or rotating shafts. They function via a sliding action.

Sleeve bearings can be relatively simple pressed-in devices used for a host of applications from guide post bushings to caster bearings. Sleeve bearings are often made of bearing bronze either sintered or cast and sometimes filled with plugs of lubricant such as graphite as with the bearings at left. Various plastics are also popular for sleeve bearings. Sleeve bearings are offered in two primary styles, a plain cylindrical version which is pressed flush into a component, and a flanged style which stands proud of the component into which it is pressed and provides a bearing surface for axial loads. Some manufacturers refer to the former type as "sleeve" bearings and the latter type as "flanged" bearings.

Flange bearings to Flange bearings support a shaft that runs perpendicular to a bearing's mounting surface. The flange (or rim) of the bearing can also be used as a locating mechanism to hold a sleeve bearing in place. Flange bearings reduce friction between surfaces in rotary and linear movements.

Mounted bearings to achieve an ideal fit, mounted bearings must be designed exactly to spec. Mounted bearings that fit too loosely can creep or slip on a shaft. Or if the press fit is too tight, free movement can be impeded. To eliminate this concern, plastic plane mounted bearings are available in pillow-block or flange housings, in forms ranging from 2-4 holes.

Thrust bearings to these plane bearings are designed with a simple washer to prevent metal-to-metal contact in a thrust load application. Plastic thrust bearings are thin, easy to install and self-lubricating to reduce maintenance costs.

Spherical bearings to Spherical bearings rotate from two directions to compensate for any shaft misalignment. They are typically called on to support a rotating shaft that calls for both rotational and angular movement.

They are distinct from spherical roller bearings, which are rolling element bearings addressed in the family Bearings. Generally, for spherical bearings, the spherical inner race rotates angularly within limits in the outer race while grease, PTFE, etc. provides a lubricating layer between sliding surfaces. In very demanding applications such as aerospace control linkages, small bearing balls roll between the inner and outer races, making for very low friction motion. Spherical bearings are not intended to handle rotation per se, though often as linkages move through their range the connected parts rotate and move angularly with respect to each other. Perhaps the most common application of spherical bearing is in rod ends.

A ball bearing is a type of rolling-element bearing that uses balls to maintain the separation between the moving parts of the bearings - the inner and outer part of the bearings.

The function of a ball bearing is to connect two machine members that move relative to one another in such a manner that the frictional resistance to motion is minimal. In many applications one of the members is a rotating shaft and the other a fixed housing.

There are three main parts in a ball bearing: two grooved, ringlike races, or tracks, and a number of hardened steel balls. The races are of the same width but different diameters; the smaller one, fitting inside the larger one and having a groove on its outside surface, is attached on its inside surface to one of the machine members. The larger race has a groove on it's inside surface and is attached on its outside surface to the other machine member. The balls fill the space between the two races and roll with negligible friction in the grooves. The balls are loosely restrained and separated by means of a retainer or cage.

Ball bearings can be broadly classified into the following:

Deep-Groove Ball Bearing

- Angular-contact ball bearing
- Self-aligning ball bearing
- Thrust ball bearing

Deep-groove Ball Bearings

The widely used ball bearing to support radial load is 'Deep-Groove ball bearing' or 'Conrad-bearing'.

They are primarily designed to support high radial load and moderate thrust load. They have deep raceways that are continuous (i.e. there are no openings or recesses) over all of the ring circumferences. This type of construction permits the bearings to support relatively high thrust load in either direction. In fact the thrust load capacity is about 70% of the radial load capacity. A ball bearing primarily designed to support radial load can also support high thrust load; because only few balls carry the radial load, whereas all the balls can withstand the thrust load.

The double-row deep-groove ball bearings have two rows of balls rolling in two pairs of races. They have more radial load capacity than that of single row bearings. In other words they are smaller in diameter compared to single row ball bearings for comparable radial load capacity. However, the proper load sharing between the balls mainly depends on the accuracy of manufacturing.

Angular Contact Bearings

The angular contact bearings are designed such that the centerline of contact between balls and raceways is at an angle to a plane perpendicular to the axis of rotation. This angle is called "contact angle". The angular contact ball bearing may be of single or two rows of balls. They are meant to carry radial and axial load together or only axial load depending on the magnitude of the angle of contact. The bearings having large contact angle support heavy thrust. The groove curvature radii are generally 52 to 53% of ball diameter. Angular contact single row ball bearings have high radial load and high unidirectional thrust load capacity than the deep groove ball bearings.

The contact angle is usually less than 40°.In the case of angular contact ball bearings, one side of the outer race is cut to insert balls. This permits the bearing to take the thrust load in only one direction. Therefore, single row angular contact ball bearings are generally used in pairs. In the case of double row angular contact ball bearings (duplex), the balls can be arranged 'back to back' and face to face' or 'tandem' configurations. The back to back and face to face duplex bearings can accommodate radial load and axial loads in both directions. The tandem bearings can accommodate radial load and heavy axial load in only one direction.

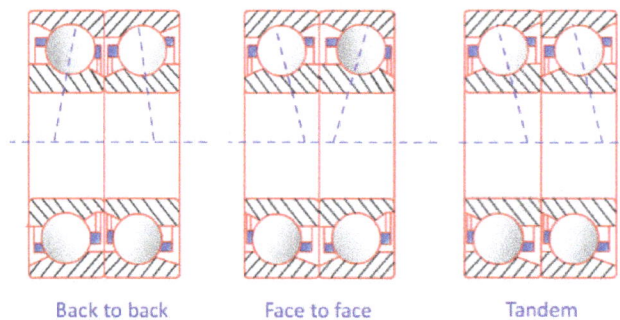

Back to back   Face to face   Tandem

Fig: Duplex angular contact ball bearings

Self Aligning Ball Bearings

For assembly of shaft and housing which cannot be made perfectly coaxial, the self- aligning ball bearings are best used. They consist of two rows of balls on a common spherical outer race. In such bearings the assembly of inner ring and balls can tilt in the outer ring. The loss of load-carrying capacity is inherent in this construction, due to non-conformity of outer raceway with the balls. This is compensated by having large number of balls in the bearings. Self-aligning ball bearings are used in top drafting rollers and main shaft of ring spinning machine.

Thrust Ball Bearings

If the contact angle of angular contact bearings exceeds 45°, it is classified as 'thrust bearing'. The maximum value this angle can assume is 90° . In such case, races are on the sideways. Such a bearing cannot take any radial load, and is used only for thrust loads. The shafts carrying bevel or worm or helical gears should be mounted with thrust bearings, except the shafts carrying honeycomb (Herringbone) gears or crossed helical gears of left- and right hands placed alternatively along the shafts.

Common Designs

There are several common designs of ball bearing, each offering various performance trade-offs. They can be made from many different materials, including: stainless steel, chrome steel, and ceramic (silicon nitride (Si_3N_4)). A hybrid ball bearing is a bearing with ceramic balls and races of metal.

Angular Contact

An angular contact ball bearing uses axially asymmetric races. An axial load passes in a straight line through the bearing, whereas a radial load takes an oblique path that acts to separate the races axially. So the angle of contact on the inner race is the same as that on the outer race. Angular contact bearings better support combined loads (loading in both the radial and axial directions) and the contact angle of the bearing should be matched to the relative proportions of each. The larger the contact angle (typically in the range 10 to 45 degrees), the higher the axial load supported, but the lower the radial load. In high speed applications, such as turbines, jet engines, and dentistry equipment, the centrifugal forces generated by the balls changes the contact angle at the inner and outer race. Ceramics such as silicon nitride are now regularly used in such applications due to their low density (40% of steel). These materials significantly reduce centrifugal force and function well in high temperature environments. They also tend to wear in a similar way to bearing steel—rather than cracking or shattering like glass or porcelain.

Most bicycles use angular-contact bearings in the headsets because the forces on these bearings are in both the radial and axial direction.

Axial

An axial or thrust ball bearing uses side-by-side races. An axial load is transmitted directly through the bearing, while a radial load is poorly supported and tends to separate the races, so that a larger radial load is likely to damage the bearing.

Deep-groove

In a deep-groove radial bearing, the race dimensions are close to the dimensions of the balls that run in it. Deep-groove bearings support higher loads than a shallower groove. Like angular contact bearings, deep-groove bearings support both radial and axial loads, but without a choice of contact angle to allow choice of relative proportion of these load capacities.

Preloaded Pairs

The above basic types of bearings are typically applied in a method of preloaded pairs, where two individual bearings are rigidly fastened along a rotating shaft to face each other. This improves the axial runout by taking up (preloading) the necessary slight clearance between the bearing balls and races. Pairing also provides an advantage of evenly distributing the loads, nearly doubling the total load capacity compared to a single bearing. Angular contact bearings are almost always used in opposing pairs: the asymmetric design of each bearing supports axial loads in only one direction, so an opposed pair is required if the application demands support in both directions. The preloading force must be designed and assembled carefully, because it deducts from the axial force capacity of the bearings, and can damage bearings if applied excessively. The pairing mechanism may simply face the bearings together directly, or separate them with a shim, bushing, or shaft feature.

Construction Types

Conrad

The Conrad-style ball bearing is named after its inventor, Robert Conrad, who was awarded British patent 12,206 in 1903 and U.S. patent 822,723 in 1906. These bearings are assembled by placing the inner ring into an eccentric position relative to the outer ring, with the two rings in contact at one point, resulting in a large gap opposite the point of contact. The balls are inserted through the gap and then evenly distributed around the bearing assembly, causing the rings to become concentric. Assembly is completed by fitting a cage to the balls to maintain their positions relative to each other. Without the cage, the balls would eventually drift out of position during operation, causing the bearing to fail. The cage carries no load and serves only to maintain ball position.

Conrad bearings have the advantage that they are able to withstand both radial and axial loads, but have the disadvantage of lower load capacity due to the limited number of balls that can be loaded into the bearing assembly. Probably the most familiar industrial ball bearing is the deep-groove Conrad style. The bearing is used in most of the mechanical industries.

Slot-fill

In a slot-fill radial bearing, the inner and outer races are notched on one face so that when the notches are aligned, balls can be slipped in the resulting slot to assemble the bearing. A slot-fill bearing has the advantage that more balls can be assembled (even allowing a full complement design), resulting in a higher radial load capacity than a Conrad bearing of the same dimensions and material type. However, a slot-fill bearing cannot carry a significant axial load, and the slots cause a discontinuity in the races that can have a small but adverse effect on strength.

Relieved Race

Relieved race ball bearings are 'relieved' as the name suggests by basically have either the OD of the inner ring reduced on one side, or the ID of the outer ring increased on one side. This allows a greater number of balls to be assembled into either the inner or outer race, and then press fit over the relief. Sometimes the outer ring will be heated to facilitate assembly. Like the slot-fill construction, relieved race construction allows a greater number of balls than Conrad construction, up to and including full complement and the extra ball count gives extra load capacity. However, a relieved race bearing can only support significant axial loads in one direction ('away from' the relieved race).

Fractured Race

Another way of fitting more balls into a radial ball bearing is by radially 'fracturing' (slicing) one of the rings all the way through, loading the balls in, re-assembling the fractured portion, and then using a pair of steel bands to hold the fractured ring sections together in alignment. Again, this allows more balls, including full ball complement, however unlike with either slot fill or relieved race constructions, it can support significant axial loading in either direction.

Rows

There are two row designs: single-row bearings and double-row bearings. Most ball bearings are a single-row design, which means there is one row of bearing balls. This design works with radial and thrust loads.

A double-row design has two rows of bearing balls. Their disadvantage is they need better alignment than single-row bearings.

Flanged

Bearings with a flange on the outer ring simplify axial location. The housing for such bearings can consist of a through-hole of uniform diameter, but the entry face of the housing (which may be either the outer or inner face) must be machined truly normal to the hole axis. However such flanges are very expensive to manufacture. A more cost effective arrangement of the bearing outer ring, with similar benefits, is a snap ring groove at either or both ends of the outside diameter. The snap ring assumes the function of a flange.

Caged

Cages are typically used to secure the balls in a Conrad-style ball bearing. In other construction types they may decrease the number of balls depending on the specific cage shape, and thus reduce the load capacity. Without cages the tangential position is stabilized by sliding of two convex surfaces on each other. With a cage the tangential position is stabilized by a sliding of a convex surface in a matched concave surface, which avoids dents in the balls and has lower friction. Caged roller bearings were invented by John Harrison in the mid-18th century as part of his work on chronographs.

Hybrid Ball Bearings using Ceramic Balls

Ceramic bearing balls can weigh up to 40% less than steel ones, depending on size and material. This reduces centrifugal loading and skidding, so hybrid ceramic bearings can operate 20% to 40%

faster than conventional bearings. This means that the outer race groove exerts less force inward against the ball as the bearing spins. This reduction in force reduces the friction and rolling resistance. The lighter balls allow the bearing to spin faster, and uses less energy to maintain its speed.

The ceramic balls are typically harder than the race. Due to wear, with time they will form a groove in the race. This is preferable to the balls wearing which would leave them with possible flat spots significantly harming performance.

While ceramic hybrid bearings use ceramic balls in place of steel ones, they are constructed with steel inner and outer rings; hence the hybrid designation. While the ceramic material itself is stronger than steel, it is also stiffer, which results in increased stresses on the rings, and hence decreased load capacity. Ceramic balls are electrically insulating, which can prevent 'arcing' failures if current should be passed through the bearing. Ceramic balls can also be effective in environments where lubrication may not be available (such as in space applications).

In some settings only a thin coating of ceramic is used over a metal ball bearing.

Fully Ceramic Bearings

These bearings make use of both ceramic balls and race. These bearings are impervious to corrosion and rarely require lubrication if at all. Due to the stiffness and hardness of the balls and race these bearings are noisy at high speeds. The stiffness of the ceramic makes these bearings brittle and liable to crack under load or impact. Because both ball and race are of similar hardness wear can lead to chipping at high speeds of both the balls and the race this can cause sparking.

Self-aligning

Wingquist developed a self-aligning ball bearing

Self-aligning ball bearings, such as the Wingquist bearing shown in the picture, are constructed with the inner ring and ball assembly contained within an outer ring that has a spherical raceway. This construction allows the bearing to tolerate a small angular misalignment resulting from shaft or housing deflections or improper mounting. The bearing was used mainly in bearing arrangements with very long shafts, such as transmission shafts in textile factories. One draw-back of the self-aligning ball bearings is a limited load rating, as the outer raceway has very low osculation (radius is much larger than ball radius). This led to the invention of the spherical roller bearing, which has a similar design, but use rollers instead of balls. Also the spherical roller thrust bearing is an invention that derives from the findings by Wingquist.

Operating Conditions

Lifespan

The calculated life for a bearing is based on the load it carries and its operating speed. The industry standard usable bearing lifespan is inversely proportional to the bearing load cubed. Nominal maximum load of a bearing is for a lifespan of 1 million rotations, which at 50 Hz (i.e., 3000 RPM) a lifespan of 5.5 is working hours. 90% of bearings of that type have at least that lifespan, and 50% of bearings have a lifespan at least 5 times as long.

The industry standard life calculation is based upon the work of Lundberg and Palmgren performed in 1947. The formula assumes the life to be limited by metal fatigue and that the life distribution can be described by a Weibull distribution. Many variations of the formula exist that include factors for material properties, lubrication, and loading. Factoring for loading may be viewed as a tacit admission that modern materials demonstrate a different relationship between load and life than Lundberg and Palmgren determined.

Failure Modes

If a bearing is not rotating, maximum load is determined by force that causes plastic deformation of elements or raceways. The indentations caused by the elements can concentrate stresses and generate cracks at the components. Maximum load for not or very slowly rotating bearings is called "static" maximum load.

Also if a bearing is not rotating, oscillating forces on the bearing can cause impact damage to the bearing race or the rolling elements, known as brinelling. A second lesser form called false brinelling occurs if the bearing only rotates across a short arc and pushes lubricant out away from the rolling elements.

For a rotating bearing, the dynamic load capacity indicates the load to which the bearing endures 1,000,000 cycles.

If a bearing is rotating, but experiences heavy load that lasts shorter than one revolution, static max load must be used in computations, since the bearing does not rotate during the maximum load.

If a sideways torque is applied to a deep groove radial bearing, an uneven force in the shape of an ellipse is applied on the outer ring by the rolling elements, concentrating in two regions on opposite sides of the outer ring. If the outer ring is not strong enough, or if it is not sufficiently braced by the supporting structure, the outer ring will deform into an oval shape from the sideways torque stress, until the gap is large enough for the rolling elements to escape. The inner ring then pops out and the bearing structurally collapses.

A sideways torque on a radial bearing also applies pressure to the cage that holds the rolling elements at equal distances, due to the rolling elements trying to all slides together at the location of highest sideways torque. If the cage collapses or breaks apart, the rolling elements group together, the inner ring loses support, and may pop out of the center.

Maximum Load

In general, maximum load on a ball bearing is proportional to outer diameter of the bearing times the width of the bearing (where width is measured in direction of axle).

Bearings have static load ratings. These are based on not exceeding a certain amount of plastic deformation in the raceway. These ratings may be exceeded by a large amount for certain applications.

Lubrication

For a bearing to operate properly, it needs to be lubricated. In most cases the lubricant is based on elastohydrodynamic effect (by oil or grease) but working at extreme temperatures dry lubricated bearings are also available.

For a bearing to have its nominal lifespan at its nominal maximum load, it must be lubricated with a lubricant (oil or grease) that has at least the minimum dynamic viscosity (usually denoted with the Greek letter ν) recommended for that bearing.

The recommended dynamic viscosity is inversely proportional to diameter of bearing.

The recommended dynamic viscosity decreases with rotating frequency. As a rough indication: for less than 3000 RPM, recommended viscosity increases with factor 6 for a factor 10 decrease in speed, and for more than 3000 RPM, recommended viscosity decreases with factor 3 for a factor 10 increase in speed.

For a bearing where average of outer diameter of bearing and diameter of axle hole is 50 mm, and that is rotating at 3000 RPM, recommended dynamic viscosity is 12 mm²/s.

Note that dynamic viscosity of oil varies strongly with temperature: a temperature increase of 50–70 °C causes the viscosity to decrease by factor 10.

If the viscosity of lubricant is higher than recommended, lifespan of bearing increases, roughly proportional to square root of viscosity. If the viscosity of the lubricant is lower than recommended, the lifespan of the bearing decreases and by how much depends on which type of oil being used. For oils with EP ('extreme pressure') additives, the lifespan is proportional to the square root of dynamic viscosity, just as it was for too high viscosity, while for ordinary oils lifespan is proportional to the square of the viscosity if a lower-than-recommended viscosity is used.

Lubrication can be done with grease, which has advantages that grease is normally held within the bearing releasing the lubricant oil as it is compressed by the balls. It provides a protective barrier for the bearing metal from the environment, but has disadvantages that this grease must be replaced periodically, and maximum load of bearing decreases (because if bearing gets too warm, grease melts and runs out of bearing). Time between grease replacements decreases very strongly with diameter of bearing: for a 40 mm bearing, grease should be replaced every 5000 working hours, while for a 100 mm bearing it should be replaced every 500 working hours.

Lubrication can also be done with oil, which has advantage of higher maximum load, but needs some way to keep oil in bearing, as it normally tends to run out of it. For oil lubrication it is recommended that for applications where oil does not become warmer than 50 °C, oil should be replaced once a year, while for applications where oil does not become warmer than 100 °C, oil should be replaced 4 times per year. For car engines, oil becomes 100 °C but the engine has an oil filter to maintain oil quality; therefore, the oil is usually changed less frequently than the oil in bearings.

Direction of Load

Most bearings are meant for supporting loads perpendicular to axle ("radial loads"). Whether they can also bear axial loads, and if so, how much, depends on the type of bearing. Thrust bearings (commonly found on lazy susans) are specifically designed for axial loads.

For single-row deep-groove ball bearings, SKF's documentation says that maximum axial load is circa 50% of maximum radial load, but it also says that "light" and/or "small" bearings can take axial loads that are 25% of maximum radial load.

For single-row edge-contact ball bearings, axial load can be about 2 times max radial load, and for cone-bearings maximum axial load is between 1 and 2 times maximum radial load.

Often Conrad-style ball bearings will exhibit contact ellipse truncation under axial load. That means that either the ID of the outer ring is large enough, or the OD of the inner ring is small enough, so as to reduce the area of contact between the balls and raceway. When this is the case, it can significantly increase the stresses in the bearing, often invalidating general rules of thumb regarding relationships between radial and axial load capacity. With construction types other than Conrad, one can further decrease the outer ring ID and increase the inner ring OD to guard against this.

If both axial and radial loads are present, they can be added vectorially, to result in the total load on bearing, which in combination with nominal maximum load can be used to predict lifespan. However, in order to correctly predict the rating life of ball bearings the ISO/TS 16281 should be used with the help of a calculation software.

Avoiding Undesirable Axial Load

The part of a bearing that rotates (either axle hole or outer circumference) must be fixed, while for a part that does not rotate this is not necessary (so it can be allowed to slide). If a bearing is loaded axially, both sides must be fixed.

If an axle has two bearings, and temperature varies, axle shrinks or expands, therefore it is not admissible for both bearings to be fixed on both their sides, since expansion of axle would exert axial forces that would destroy these bearings. Therefore, at least one of bearings must be able to slide.

A 'freely sliding fit' is one where there is at least a 4 μm clearance, presumably because surface-roughness of a surface made on a lathe is normally between 1.6 and 3.2 μm.

Fit

Bearings can withstand their maximum load only if the mating parts are properly sized. Bearing manufacturers supply tolerances for the fit of the shaft and the housing so that this can be achieved. The material and hardness may also be specified.

Fittings that are not allowed to slip are made to diameters that prevent slipping and consequently the mating surfaces cannot be brought into position without force. For small bearings this is best done with a press because tapping with a hammer damages both bearing and shaft, while for large bearings the necessary forces are so great that there is no alternative to heating one part before fitting, so that thermal expansion allows a temporary sliding fit.

Avoiding Torsional Loads

If a shaft is supported by two bearings, and the center-lines of rotation of these bearings are not the same, then large forces are exerted on the bearing that may destroy it. Some very small amount of misalignment is acceptable, and how much depends on type of bearing. For bearings that are specifically made to be 'self-aligning', acceptable misalignment is between 1.5 and 3 degrees of arc. Bearings that are not designed to be self-aligning can accept misalignment of only 2–10 minutes of arc.

Applications

In general, ball bearings are used in most applications that involve moving parts. Some of these applications have specific features and requirements:

Hard drive bearings used to be highly spherical, and were said to be the best spherical manufactured shapes, but this is no longer true, and more and more are being replaced with fluid bearings.

German ball bearing factories were often a target of allied aerial bombings during World War II; such was the importance of the ball bearing to the German war industry.

In horology, the company Jean Lassale designed a watch movement that used ball bearings to reduce the thickness of the movement. Using 0.20 mm balls, the Calibre 1200 was only 1.2 mm thick, which still is the thinnest mechanical watch movement.

Aerospace bearings are used in many applications on commercial, private and military aircraft including pulleys, gearboxes and jet engine shafts. Materials include M50 tool steel (AMS6491), Carbon chrome steel (AMS6444), the corrosion resistant AMS5930, 440C stainless steel, silicon nitride (ceramic) and titanium carbide-coated 440C.

A skateboard wheel contains two bearings, which are subject to both axial and radial time-varying loads. Most commonly bearing 608-2Z is used (a deep groove ball bearing from series 60 with 8 mm bore diameter).

Yo-Yos, there are ball bearings in the center of many new, ranging from beginner to professional or competition grade, Yo-Yos.

Many fidget spinner toys use multiple ball bearings to add weight, and to allow the toy to spin.

Roller Bearing

Roller Bearings are a type of rolling-element bearing that uses cylinders (rollers) to maintain the separation between the moving parts of the bearing (as opposed to using balls as the rolling element). The purpose of a roller bearing is to reduce rotational friction and support radial and axial loads. Compared to ball bearings, roller bearings can support heavy radial loads and limited axial loads (parallel to the shaft). They can operate at moderate to high speeds (although maximum speeds are typically below the highest speeds of ball bearings). The lubrication method must carefully considered during the design phase when using roller bearings.

There are five main types of roller bearings:

Cylindrical Roller Bearings have high radial-load capacity and moderate thrust loads. They contain rollers which are cylindrically-shaped, but crowned or end-relieved to reduce stress concentrations. Cylindrical roller bearings are similar in design to needle roller bearings but the dimensions of diameter and roller length are closer in magnitude.

Spherical Roller Bearings are self-aligning, double row, and combination radial and thrust bearings. They use a spherical or crowned roller as the rolling element.

Tapered Roller Bearings consist of an inner ring (cone), an outer ring (cup), a cage and rollers, which are profiled to distribute the load evenly across the roller. During operation, tapered roller bearings create a line contact between the raceway and rolling element, distributing loads across a larger area.

Needle Roller Bearings are a type of cylindrical roller bearing where the length of the roller is much larger than then the diameter. Needle roller bearings are designed for radial load applications where a low profile is desired.

Thrust Bearings are designed for pure thrust loads, and can handle little or no radial load. Roller thrust bearings use rollers similar to other types of roller bearings.

Components

Radial type roller bearings (cylindrical, tapered, spherical, and needle) consist of four basic components, an inner ring, an outer ring, rollers, and a cage (roller retainer). Under normal operating conditions, bearing rings and rollers carry the load while the cage spaces and retains the rollers on the cone.

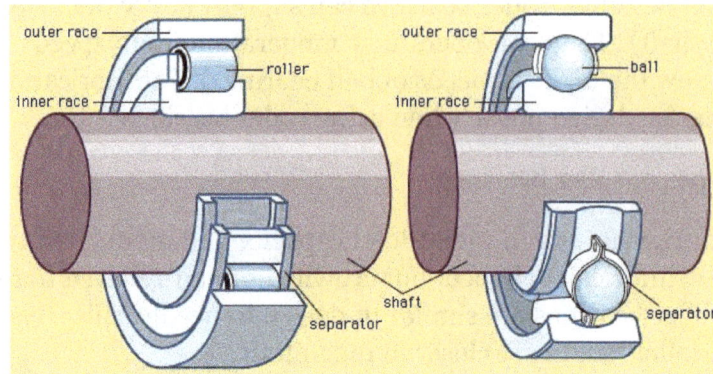

Comparison of Cylindrical Roller Bearing and Ball Bearing Components

Roller thrust bearings are designed to carry pure thrust loads. Like radial roller bearings, roller thrust bearings also consist of two rings, rollers, and a cage (roller retainer). However, instead of an inner and outer ring concentric to the axis of rotation, they have two rings or thrust washers on either side of the roller.

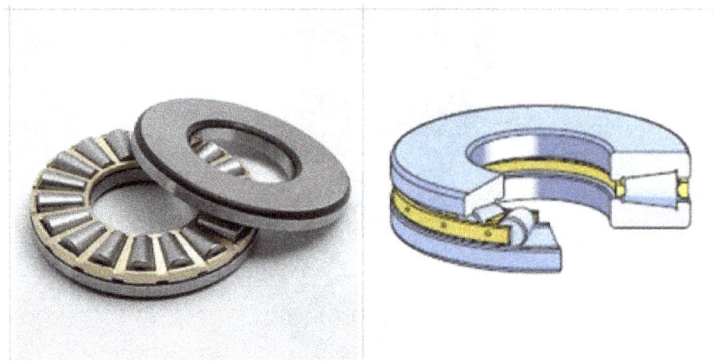

Tapered roller thrust bearings

Specifications

Key specifications for roller bearings include dimensions, rated speed abd various rated load types.

Dimensions

Important dimensions to consider when specifying bearings include:

Bore: The bearing industry uses a standard number system for radial bearings with metric diameter bores. For bore sizes 04 and up multiply by 5 to obtain the bore in millimeters. If the bore is a hex this refers to the dimension across the flats. If the bore is tapered this refers to the smaller diameter.

Outside diameter: The outside diameter of the bearing includes the housing if a housed unit, but excludes the flange if a flanged bearing. The outer ring width is the overall width of the outside of the bearing.

Overall width: The overall width of the bearing or bearing assembly includes the locking collar, if present.

Operating Specifications

Important operating specifications to consider when searching for bearings include rated speed, dynamic axial or thrust load, and dynamic radial load.

- The rated speed for a bearing running with grease lubrication is lower than a bearing with oil lubrication.

- The static axial or thrust load is the maximum load a bearing can endure parallel to the axis of rotation without excessive, permanent deformation.

- The static radial load is the maximum radial load a bearing can endure without excessive permanent deformation.

- The dynamic axial or thrust load is the calculated constant axial load, which a group of identical bearings with stationary outer rings can theoretically endure for a rating life of 1 million revolutions of the inner ring.

- The dynamic radial load is the calculated constant radial load, which a group of identical bearings with stationary outer rings can theoretically endure for a rating life of 1 million revolutions of the inner ring.

Features

- Flanged: The bearing has a flange for mounting or locating.

- Spherical outside diameter: The bearing is tolerant of misalignment and has a greater load capacity than internal self-alignment, but requires more radial space.

Materials

Materials used for roller bearings are typically alloy or low carbon steels. Some applications require case or thoroughly hardened high carbon bearing quality steel. Depending on the size of the bearing to be produced, appropriate quantities of alloying elements are added to the steel melt to assure optimum properties in the finished product.

When low carbon carburized grades of steel are used, carbon is often introduced into the surfaces of the roller bearing components after machining, to a depth sufficient to produce a hardened case that will sustain bearing loads. The carbon and alloys added earlier ensure a proper combination of hard fatigue resistant case and a tough ductile core. High carbon grades of steel do not require carburizing and are either case-hardened, normally by induction heating, or through-hardened by conventional heating methods.

Another benefit derived from case carburizing tapered roller bearings is the development of residual compressive stresses in the surface layers. These residual stresses retard the propagation of fatigue cracks that initiate close to the bearing raceway and roller surfaces. This helps to improve the bending fatigue resistance at the large rib undercut and ensure the ability to endure heavy shock loads without damage. The hardened case of roller bearing components provides fatigue resistance and the ductile core provides toughness for the roller bearing.

Bearing Setting

The bearing setting of a roller bearing is defined as the specific amount of either endplay or preload.

Tapered roller bearings have the inherent advantage of being adjustable; therefore, they can be set to approach optimum performance in almost any application. They may be set manually, supplied as a preset assembly, or set by using an automated setting technique.

Automated roller bearing setting techniques offer many advantages such as reduced setting time, decreased assembly cost, and consistent and reliable setting with minimal skill requirements. They may be applied to both assembly line and field repair applications.

Selection

Selection of roller bearings consists of two steps. First, one must determine the desired bearing life and then select the roller bearing with a sufficient basic dynamic load rating to meet that life requirement.

Applications

Roller bearings are used for heavy-duty moderate-speed applications. Potential applications for spherical and cylindrical roller bearings include power generation, oil field, mining and aggregate processing, wind turbines, gear drives, rolling mills. Single-row tapered roller bearings are used in such applications as machine tool spindles, gear reduction units, automotive transaxles, transmissions, vehicle front wheels, differential and pinion configurations, conveyor rolls, machine tool spindles, and trailer wheels.

Air Bearing

Air bearings use a thin film of pressurized air to support a load, the same way the puck on an air hockey table "floats" on air. This type of bearing is called a "fluid film" bearing. Fluid film bearings have no solid-to-solid contact under typical running conditions; instead, a film of lubricating fluid (in our case pressurized air) forms a layer between the solid machine elements and serves to transfer forces from one to the other. To compare this with ball bearings, in ball bearings the balls are in constant contact with and form a solid bridge between the machine elements.

In an air bearing, the balls are replaced by a cushion of air. Perhaps one of the most familiar applications of an air bearing is a hovercraft. Large fans blow air underneath the hovercraft and the air is prevented from escaping by a flexible rubber 'skirt'. The high air pressure generated under the hovercraft is capable of supporting its weight and it therefore floats on this cushion of air. The large cushion of air not only acts to support its weight but also as a soft spring which allows it to float smoothly over the rough surfaces on land or in the water.

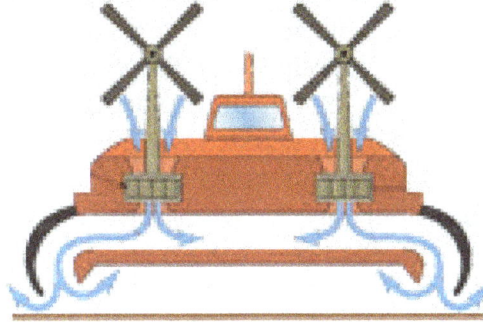

It is possible to use this same principle to act as a bearing for a rotating shaft. High pressure air is fed into the gap between the rotating shaft and a stationary bearing. The gap is extremely small (around 1/100th of a millimetre) allowing the air pressure to be maintained within the gap. This small gap also significantly reduces the 'springiness' of the air cushion resulting in a shaft that is very accurately located i.e. has low dynamic run out. The shaft can then rotate freely as there is very little friction and the air pressure will ensure the shaft does not come in contact with the stationary bearing surfaces.

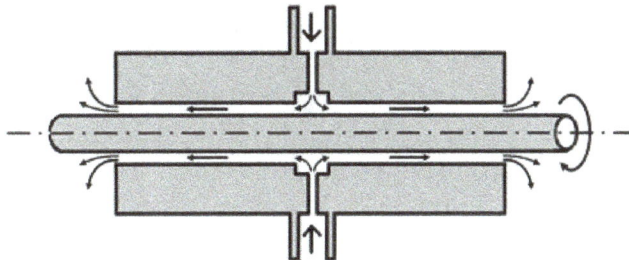

Classes of Air Bearings

Air bearings fall into two main classes, which indeed correspond to the two major representative classes in liquid film lubricated bearings: aerostatic bearings, which require a feed of pressurized air for their operation, and aerodynamic bearings, which generate their own internal pressure differentials. The latter generate this pressure by the action of simultaneously shearing and squeezing the environmental gas between the surfaces in relative motion, whereas the former require an external pump to produce the pressure. Both these types of bearing can be employed to sustain either axial or radial loads in rotating systems (or direct loads in linear configurations), or to combine the functions in a single member. A bearing can operate either entirely aerostatically or entirely aerodynamically throughout its operating speed range, or the bearing can start its movement in one mode of operation and transfer to the other as the speed changes, or it can operate with a combination of both aerostatic and aerodynamic pressure generation. Under certain circumstances, moreover, it may be necessary to supply externally pressurized gas to a self-acting bearing to prevent instabilities from arising. The flow regime in either type is usually laminar, but turbulent flow conditions can occur.

In general, the externally pressurized bearings (aerostatic), for many purposes suffer an inherent disadvantage through requiring a pressure source and an exhaust sink; but they can be made to relaxed manufacturing tolerances and can provide support at low speeds, and sustain intermittent or stationary loads. The self-acting bearings (aerodynamic), on the other hand, are able only to support a few pounds per square inch of bearing area depending upon the speed, require careful manufacture and alignment and are only suitable for bearings whose surfaces are always moving when under load. However, such bearings do not require auxiliary equipment and there is no problem of disposing of the exhaust gas or arranging for appropriate pressure control of the working compartment. The pressure gas for the externally pressurized type may, however, offer special advantages in particular circumstances. For example, a pressure-fed bearing can work in a dust-laden atmosphere because the exhaust air prevents the entry of solid particles from the environment.

Despite some limitations of load carrying capacity compared to other mechanical or rolling element bearings there are numerous advantages, and some of them exclusive, that permit machine operation in conditions which would otherwise be quite impracticable.

Exclusive Advantages of Air Bearings

The advantages which air lubrication can offer stem from the properties of gases: first, they are chemically stable over a wide temperature range and second they have inherently low viscosities. However, even in this category it is often practical to design an alternative device by a fundamentally different approach to the same job without air bearings, and so it is really not feasible to draw a hard line between exclusive advantages and advantages of degree. Nevertheless, it is believed that this way of thinking of the advantages available does avoid some confusion of thought since it focuses attention upon aspects of a proposed application in which the gas really may be doing something which no other lubricant could possibly do for fundamental reasons and those in which there is merely competition in regard to convenience or cost.

Low Friction

Of the more important exclusive advantages which are offered by air lubrication are those cases where the low viscosity of gases as compared with liquids can be exploited to special benefit. Particularly straightforward examples of this class of application are those which occur in near-static apparatus such as gimbal support, dynamometers, wind-tunnel balances and other specialized mechanical instruments which benefit from the extremely low static frictional torque which externally pressurized bearings can offer. The use of a gas permits a torque orders of magnitude smaller than could be achieved by liquids, but perhaps in practice often of more importance is the fact that a low-torque bearing with an appropriately large operating clearance can be made in a very simple and clean fashion using gas lubrication. Air is usually employed, since the exhaust from the bearing can be released to the surroundings and quite large flow rates can be employed.

In semiconductor machine tools, high-speed, acceleration and damping are key to product throughput. The use of air bearings in such an application has found widespread use. Experimental high speed linear slides of a composite lightweight structure have operated at 14g acceleration for thousands of hours or repeated because of the low friction aspects of air bearings. A mechanical rolling element-type bearing would never be able to satisfy such a requirement. This low friction also finds uses in torque measuring equipment, dynamic balancing machinery, semiconductor positioning systems, micro or zero gravity trajectory simulators and other instruments requiring near-static conditions.

High Accuracy

The high accuracy of motion that can be obtained with air bearings is equally important in some applications. Considerable differences in motion accuracy exist between rolling element bearing supports and air bearing supports. In linear slides, for example, rolling element bearings witness noise error (or rumbling) due to the ways' surface roughness and/or eccentric rotation of the rollers or balls.

On the contrary, air bearings do not suffer from this difficulty. The reason for this lies in the absence of surface contact between the bearing parts and the averaging action of the air film over the various local surface irregularities present in the machined surfaces. Even the finest of rolling element bearings are orders of the magnitude less accurate than air bearings. In rotating air bearings, this effect produces high orders of rotational accuracy and smoothness of travel. Typical T.I.R. for air bearing spindles are less than 1.0 µinch. For linear slides, pitch, roll and yaw errors of much less than a fraction of an arc second are attainable and straightness of travel errors on the order of nanometers have been achieved.

High Stiffness

At zero speed, air bearings provide considerably high stiffness characteristics. This same effect is seen at zero or low loads. For properly designed and manufactured aerostatic bearings, it is not uncommon to measure stiffness on the order of several million pounds per inch.

Zero Wear

The advantage of zero wear can be seen greatly in externally pressurized or aerostatic bearings and to some large degree in self-acting or aerodynamic bearings. Although some properly designed rolling element bearings can achieve practical wear rates, none can match the zero wear characteristic of aerostatic bearings. With aerodynamic bearings, starting and stopping causes some rubbing within the bearing clearance, but this can be alleviated by introducing a pulse of air just as the bearing begins translation. Furthermore, as compared with rolling element bearings, air bearings do not suffer from increased wear rates as the speed or load is increased. With proper care and maintenance, infinite life can be expected from air bearings.

Note on crash resistance: allowing a bearing to crash or be overloaded to the grounded state should never be a design feature. Crashed bearings are a sign of a much wider system problem and should be corrected regardless of the type of bearing used.

Contamination

Gas lubrication has found a place of particular importance in circumstances where it is necessary to keep the environment free from contamination by conventional lubricants. Such situations arise in semiconductor wafer handling systems. In these situations, it may be costly or impractical to manufacture a system which can effectively seal off contaminants from oil lubricants used in roller slides. The externally pressurized air bearing lends itself well to harsh environments where liquids, dust and contaminants are present. The air bearing's great resilience stems from the fact that with positive pressure existing inside the bearing, all foreign matter is repelled away from the critical bearing surfaces. Externally pressurized bearings could operate while completely submerged in a liquid. Unlike some rolling element bearing supports that require periodic maintenance, cleaning, the addition of oil lubricants and sometimes the replacement or re-surfacing of guide ways, the air bearing's self-cleaning nature allows it to be virtually maintenance-free.

Wide Temperature Range

Perhaps the most exclusive quality of gases as lubricants is their potential for operation over extremely wide ranges of temperature. Indeed, it is the invulnerability of the solid components of the machine, not that of the lubricant, which will set performance limits when simple gases are used for high temperature lubricated applications, although at the lower end of the temperature scale condensation of the gas may become a limitation. Complex gases on the other hand will have decomposition limitations at the upper end of their usable temperature range. No difficulty is seen, for example, at the hot end end of the scale, in operating the bearings of small steam turbines or circulators upon superheated steam, and, at the cold end, gases approaching their liquefaction temperatures could be employed to lubricate the bearings of, for instance, gas liquefying turbines. In both examples considerable simplification of design could thus be achieved in some situations. It is noted that whereas with liquids bearing performance falls off with rise of temperature due to fall in viscosity, in the case of gases, the load-carrying performance will in general improve due to a rise in viscosity.

Externally pressurized ceramic bearings were operated at temperatures of up to 1,500°F (800°C) at speeds of up to 65,000 rev/min. Low temperature applications of air bearings have been largely confined to various types of expansion turbine gas liquefiers and to refrigeration plant. A small high-speed expansion turbine for liquefying helium can operate at 350,000 rev/min. This unit employs bearings which are lubricated by helium gas at a temperature between 50° and 13°K, and the output of liquid helium is maintained at a pre-selected temperature of between 3° and 4°K to an accuracy of 0.05°.

Spiral Groove Bearing

A spiral-groove bearing is a self-acting bearing with a repetitive pattern of grooves in one of its two bearing surfaces. When these surfaces move in the right direction in relation to one another, the pumping action of the grooves produces an overpressure in the lubricant between them. Because of this overpressure the lubricant keeps the moving surfaces apart and a force is transferred.

In many applications in professional equipment the spiral-groove bearings are often lubricated with air. This does not contaminate the immediate surroundings of the bearing, it can operate at extremes of temperature and the speed of the shaft can be very high.

Operation of spiral-groove Bearings

To illustrate the operation of the bearing let us first examine a simple model consisting of a plain bearing surface in motion and, parallel to it and underneath it, a stationary bearing surface with a pattern of parallel grooves at an angle a to the velocity vector. The presence of the grooves produces an overpressure between the bearing surfaces because the lubricant, the 'viscous medium', is 'dragged' through narrowing gaps (the 'wedge effect'). Because the grooves are at an acute angle, lubricant is also pumped along the grooves so that the pressure increases in this direction. The mean pressure difference Sp between the two ends of the groove pattern (y = d and y = o) depends on a number of quantities

$$\Delta p = \frac{nudg}{h_r^2}$$

where n is the viscosity of the lubricant, v the velocity of the upper bearing surface, h_r, the distance between the upper bearing surface and the ridges in the lower bearing surface, and g is a function of the groove angle a, the ratio fJ of the ridge width b, to the groove width bg, and the ratio b of n, to the groove depth hg. Calculations show that for grooves of rectangular cross-section the function g has a maximum value (about 0.09) when a = 16°, B == I and b = 0.4.

In most cases both theory and practical applications relate to grooves of rectangular cross-section. This is not essential for the bearing action: any cross-section is in principle possible and can be used in practice provided it has been dimensioned correctly. In fig. 3 the function g is plotted against o for a triangular groove cross-section for different values of a. The envelope curves have also been drawn for rectangular, circular and triangular cross-sections. These curves show that the greatest pressure is produced with a rectangular cross-section and also that the pressure is not much lower for the other two cross-sections.

The results of the simple model with parallel grooves can be used to determine the best shape of groove for a thrust bearing . When the shaft is rotating in the correct direction with respect to the groove pattern the grooves pump lubricant towards the centre of the bearing. This bearing can be thought of as being split up into a large number of elements each resembling the model described above. Each 'ring' at an arbitrary distance r from the centre with a radial width Sr functions as a pump. The pumping action developing the pressure is at its most effective when the pressure difference across! lris as large as possible. This means that all the grooves should be at the same angle to the local velocity vector. This requirement is satisfied if the grooves have the shape of a logarithmic spiral. Similarly it may be deduced that in journal bearings the grooves should take the form of a helix.

It will be clear that these pumps will produce a pressure in the lubricant between the rotating and stationary bearing surfaces. The load-carrying capacity of the bearing can be calculated by integrating the pressure over the entire bearing surface. Allowance must be made here for the finite number of grooves.

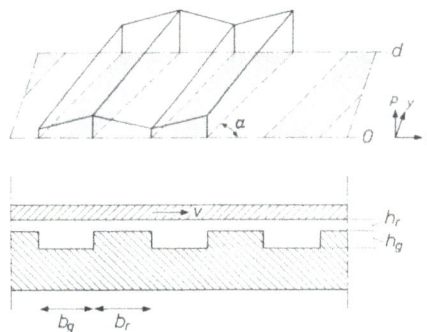

A diagram showing the distribution of the pressure

A diagram showing the distribution of the pressure p in the lubricant, the 'viscous medium', between a plain bearing surface moving at a velocity v and a stationary bearing surface parallel to it, with grooves that have a rectangular cross-section and are at an angle a to the velocity vector. h, height of gap above the ridges, hg groove depth, b, ridge width, bg groove width. There is a constant pressure gradient in the direction of the grooves.

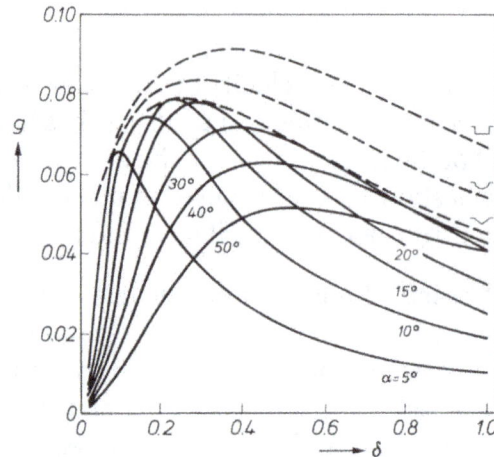

Figure: The function g for f3 = 1 plotted against o for different values of a. The solid lines refer to grooves of triangular cross-section. The dashed lines represent the envelope curves for rectangular, circular and triangular groove cross-sections. The highest value (about 0.09) is obtained for grooves of rectangular cross-section when $\alpha \approx 16^0$ $and\, \delta \approx 0.4$.

The three-dimensional representation of the pressure distribution in a spiral-groove thrust bearing that we gave in an earlier issue of this journal applies only to a model with an infinite number of grooves. In an actual practical bearing the pressure ripples that we have shown in figure appear on the outside of this 'pressure hill' as a kind of modulation. The pattern of the pressure distribution in a thrust bearing with k grooves is shaped like an inverted pudding basin with k ribs on its surface.

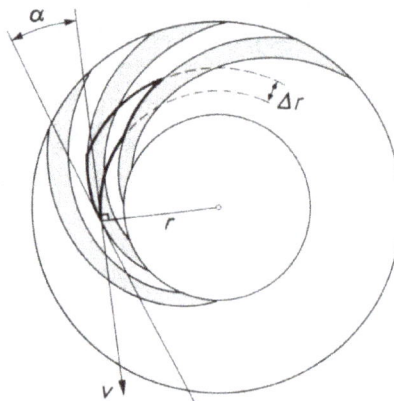

Figure: Part of a spiral-groove pattern for a thrust bearing.

Each 'ring' of width Sr, at a distance r from the center, pumps lubricant towards the center. The pumping action is strongest when the grooves have the shape of a logarithmic spiral, so that the angle a between the local velocity vector v and the tangent to the groove is constant over the entire grooved surface.

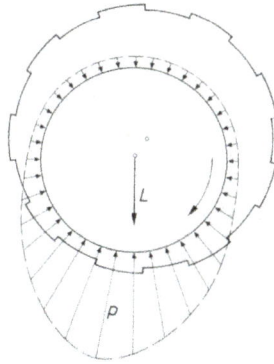

Figure: Diagram of the distribution of the pressure p in a spiralgroove journal bearing in which the shaft runs eccentrically because of the load. L load vector. Only the overpressure in the constricted part of the bearing gap gives the bearing a radial load-carrying capacity.

The grooves in a spiral-groove journal bearing also build up a pressure between the bearing surfaces. However, when the shaft is concentric with the bearing bush this does not produce a radial load-carrying capacity. Only if the shaft is in an eccentric position is a radialload-carrying capacity produced by the overpressure created in the narrowing part of the bearing gap. The grooves ensure that the lubricant is kept inside the bearing.

The Groove Pattern

Figure gives diagrams of three different groove patterns for a spiral-groove thrust bearing. For a 'blind' thrust bearing a completely grooved bearing surface is used. In this bearing there is no net flow of lubricant from the outside to the center. If the same pattern is used for a thrust bearing with a straight through shaft, the pumping action cannot maintain a sufficiently high pressure because the lubricant leaks away too much in the immediate vicinity of the shaft. This leakage can be restricted by reducing the radial length of the groove pattern the plain part then acts as a resistance to the flow. If a 'herringbone pattern' is used, the leakage around a straight-through shaft can be eliminated completely because the pumping actions of the two sections of the herringbone are in opposing directions.

A herringbone pattern is also suitable for journal bearings. The groove pattern can extend over the entire length of the bearing or merely over a part of it, with a part of the bearing surface between the two sections of the herringbone left plain; see figure. The two sections of the herringbone do not have to be of the same length. When grease is used as the lubricant, an asymmetrical herringbone pattern with a completely grooved bearing surface is preferred. An asymmetrical pattern of this kind ensures that any spare grease remains on the side of the bearing that the designer intended, because the longest section of the herringbone can deliver a higher pump pressure than the shortest. When the bearing surface is completely grooved it is easier to prevent the occurrence of under pressure areas in the grease layer that might cause cavitation. Air bubbles can accumulate in such areas, so that the space between the bearing surfaces becomes incompletely filled with grease.

For the operation of a spiral-groove bearing it does not matter whether the groove pattern is in the rotating bearing surface or in the stationary one. It does make a difference, however, whether the pattern rotates with respect to the load vector or not. Figure shows four different situations for a journal bearing with a non-rotating load vector. In two cases the grooved surface rotates and in the other two the plain surface rotates. Situations where the load vector rotates are also possible, and their behavior can also be predicted theoretically.

The 'Viscous Medium'

The use of air as a lubricant has the advantage that the bearing is always 'immersed' in a bath of lubricant. Air-lubricated spiral-groove bearings are highly suitable for use at high temperatures and very high speeds. A disadvantage of air is its low viscosity; this often makes it necessary to use bearings with impracticably large diameters and a very small clearance.

Also, air offers no protection from corrosion or from damage when the bearing surfaces touch on starting and stopping. For bearings that do not operate at such high speeds, oil is therefore the usual lubricant. A disadvantage of oil is that it can easily leak away when the bearing is at rest, because of the action of gravity. Extra precautions then have to be taken to maintain continuous lubrication and prevent the environment of the bearing from being contaminated. We took the unusual step of selecting grease as the lubricant instead of oil. So far, the spiral-groove bearing is the only self-acting bearing that permits this option without the need for re-lubrication. When grease is used, much less leakage occurs at rest, because the grease does not start to flow until the shear stress exerted on it exceeds a certain limiting value, the yield point. In an operating spiral-groove bearing grease behaves almost like oil.

Figure: Herringbone pattern for a completely grooved bearing surface (left) and for a partially grooved bearing surface (right) in a spiral-groove journal bearing.

The two groove patterns are symmetrical: the two sections of the herringbone that pump lubricant to the center are of the same length. When grease is used as the lubricant an asymmetrical pattern is preferred.

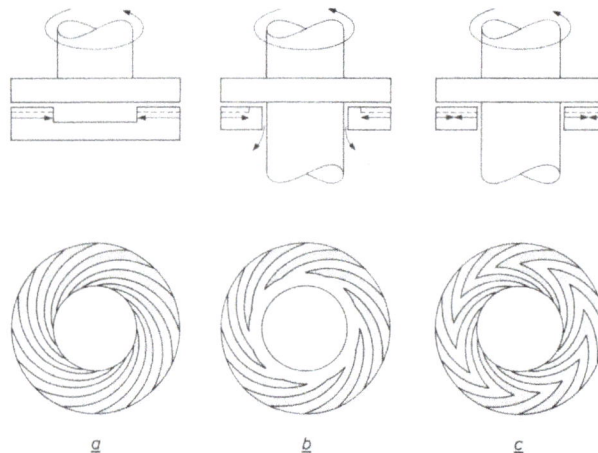

<u>a</u>   <u>b</u>   <u>c</u>

Figure: Three types of spiral-groove bearing

Three types of spiral-groove bearing for a thrust-bearing system. a) Completely grooved blind thrust bearing. b) Bearing for a straight-through shaft, with shorter grooves. c) Herringbone bearing for a straight-through shaft. The straight arrows indicate the direction in which the lubricant is pumped. The downward-pointing arrows indicate the places where the lubricant leaks out.

The operation of the bearing is affected by flow movements of the lubricant. We can consider the net flow through the bearing ('macro-flow'), or the flow in the grooves and above the ridges ('micro-flow').

If sufficient lubricant is fed into the bearing from outside, a journal bearing with an asymmetrical groove pattern will itself create macro-flow because one section of the herringbone has a stronger pumping action than the other. The occurrence of macro-flow is sometimes desirable since it may help to cool the bearing. For some bearing geometries and operating conditions this macro-flow may be stronger than the cooling flow produced by an external pressure difference. Macro-flow has practically no effect on the load-carrying capacity or the stiffness of a journal bearing, but does have an effect on the stability and the frequency of the 'half-speed whirl' generated by the bearing .

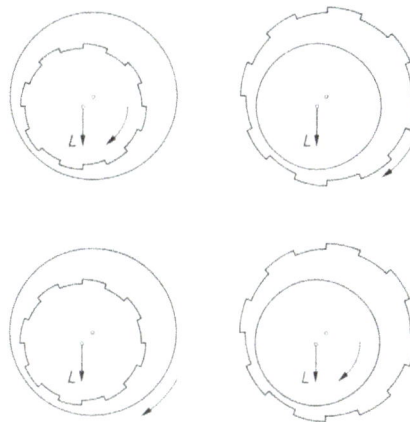

Figure: Diagram showing four situations for a journal bearing with a non-rotating load vector L; grooves in shaft or bearing bush; shaft or bearing bush rotates.

The bearing action depends only on whether the grooves rotate with respect to the load vector (the upper two situations) or are stationary with respect to the load vector (the lower two situations).

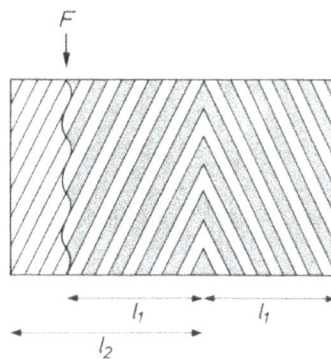

Figure: Asymmetrical herringbone pattern of a completely grooved bearing surface in an oil- or grease-lubricated spiral-groove journal bearing.

When there is a limited amount of lubricant a free boundary F arises in the left-hand section (length '2) at a distance from the center that is equal to the length 'I of the right-hand section so that the two sections pump equally strongly towards the center.

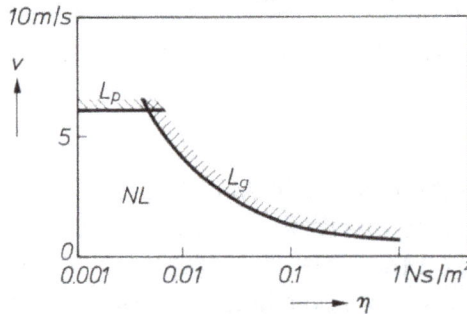

Figure: Diagram showing leakage of lubricant in a spiral-groove journal bearing as a function of the circumferential velocity vand the viscosity 17of the lubricant. NL no leakage.

Lp leakage via the plain bearing surface; occurs only at a high circumferential velocity if the lubricant has a low viscosity. Lg leakage via the grooved bearing surface; occurs even at a low circumferential velocity if the viscosity is high.

The macro-flow that arises, after starting, in a journal bearing with an asymmetrical groove pattern and with a limited quantity of lubricant stops as soon as the two sections of the herringbone have been filled to the same extent. This means that the longer section will now have a 'free boundary', which acts as a non-leaking seal because of the surface tension, provided the circumferential velocity is not too high. The velocity at which leakage commences will depend to a large extent on the viscosity of the lubricant see fig. la. If the viscosity is low, leakage in most spiral-groove journal bearings will take place only at high speeds at the plain bearing surface, whereas it will occur much sooner along the grooved bearing surface if the viscosity is high.

Even when there is no macro-flow, there will still be micro-flow in a spiral-groove bearing. The complex flow field present in the groove pattern causes a quantity of lubricant to flow over the ridges towards the outer edge, in other words out of the bearing. The same quantity of lubricant flows back from the outer edge through the grooves into the bearing.

Greases for spiral-groove Bearings

A grease for a spiral-groove bearing must have the correct viscosity and yield point. The existence of a yield point τ_y implies not only that the grease will not begin to flow until the shear stress. Exerted upon it is greater than τ_y, but also that the dynamic viscosity η depends on the shear rate $\dot{s}$:

$$\eta = \tau/\dot{s} = \tau_y/\dot{s} + \eta_\infty$$

The viscosity only approaches a constant value η_∞ at high shear rates. Strictly speaking it would be more appropriate here to speak of 'the apparent dynamic viscosity', but for convenience we shall continue to talk about 'the viscosity'. The diagram of fig. 11 illustrates the difference between a fluid with a yield point (a Bingham fluid) and a Newtonian fluid, which does not have a yield point and has a viscosity independent of the shear rate.

As well as meeting the requirements for viscosity and yield point suitable grease must also satisfy some 'other general requirements. It must, for instance, protect the bearing surfaces from any damage that might occur when they are in contact with, one another, even during rotation, and from corrosion from outside, but it must not attack the bearing surfaces or other structural materials. In many consumer articles it is extremely difficult or even impossible to relubricate the bearing once it has been assembled. Thè grease must therefore have a long life. This means that it must be mechanically and chemically stable must not oxidize, and that the base oil must not evaporate or leak away from the grease.

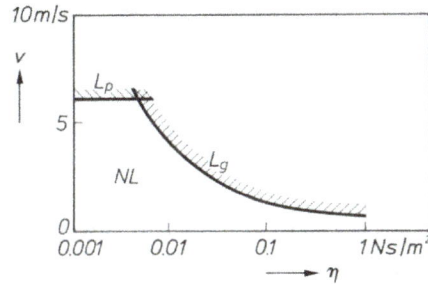

Figure: Viscosity Tl as a function of the shear rate s. For a Newtonian fluid N, Tl is constant; for a Bingham fluid-B, Tl is a decreasing function of s, which only tends to a constant value Tl cc when the shear rate is high.

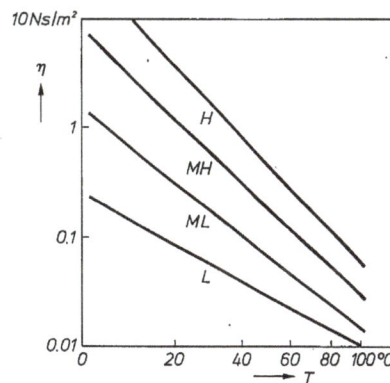

Figure: Viscosity Tl of four basic greases that we used, as a function of temperature T. The designations L, ML, MH and H stand for 'light', 'medium-light', 'medium-heavy' and 'heavy'. This series of greases will cover a wide viscosity range'.

Because of the special conditions in spiral-groove bearings it is difficult for a grease to meet all the above - requirements in full. It must, for instance, have a sufficiently high yield point to prevent leakage when at rest, yet all of the grease must start to flow as soon as rotation starts. The amount of grease used in a spiral groove bearing is usually very small, only a few tens of mm" per bearing. The grease is continuously subject to shear; the shear rate in the bearing gap is about 104 to 105 S-1. We should remember here that a required life of several thousand hours is more the rule than the exception. There is also an intensive circulation of grease (micro-flow) in the bearing so that there is always 'fresh grease' arriving at the edges of the bearing and being exposed to oxygen there. Because of these special circumstances greases for spiral groove bearings should generally be quite different in composition from greases used in ball bearings, for example.

Composition

A grease consists mainly of a base oil and a thickener; the first functions as the fluid lubricant, and the second determines the consistency.

In simple terms we could say that the thickener retains the liquid lubricant like a sponge. Another concept is to imagine the thickener as a three-dimensional network resembling a bundle of sticks to which the base oil is attached by physical and chemical bonds. When the shear stresses are small, the 'sticks' of the thickener deform elastically; the grease behaves like an elastic solid. When the stresses exceed the yield point, the 'sticks' slide about on one another; the grease then behaves like a plastic material. The sliding sticks break and begin to line up with the direction of flow so that the mixture of sticks and oil begins to behave more and more like a fluid. As the shear stress increases, the pieces of broken stick become smaller, until equilibrium is reached. The grease then begins to approximate more closely to a real fluid containing solid round particles. If the shear is stopped after a while, the thickener particles start to reorientate themselves somewhat and to recrystallize, but the yield point now has a considerably lower value than in the unworked grease.

To cover a sufficiently large range of viscosity it is not in fact necessary to stock greases of many different viscosities; in our view a range of at least four should be sufficient. The four greases should have the same type of base oil and the same thickener, so that intermediate viscosities can be obtained by mixing. The differences in viscosity then depend mainly on differences in the mean molecular weight of the base oil. Figure above shows the viscosity as a function of temperature for a range of greases that we use. The greases have been designated L ('light'), ML ('medium-light'), MH ('medium-heavy') and H ('heavy'). The relation between the temperature and the viscosity can be expressed as:

$$\eta(T) \approx k \exp [b/(T+95)]$$

where k and b are constants for a given grease, and T is the temperature in degrees Celsius. The variation with temperature increases with the mean molecular weight of the base oil. For a medium-heavy oil a rise in temperature of 10 °C reduces the viscosity by almost a half. Because it contains a relatively small amount of thickener, a suitable grease has a viscosity not much higher than that of its base oil: usually within a factor of two. The temperature dependence of the viscosity of a grease is always less than that of its base oil: as temperature increases, more base oil combines with the thickener particles or becomes attached to them to give 'thermal swelling', and the viscosity therefore decreases less sharply.

Rheological Behavior

Because re-lubrication is undesirable or sometimes even impossible the supply of grease to the bearing surfaces must be assured. In a spiral-groove journal bearing the grease is introduced into a reservoir immediately after assembly of the bearing at a point that allows it to come into contact with the wall of the reservoir and with the wall of the shaft. When the bearing runs for the first time, the grease has to be transported from the reservoir to the bearing gap. It is then forced further into the gap by the pumping action of the grooves. The distribution of the grease can be observed by using a transparent bearing, for example of polymethyl methacrylate (Perspex); see figure below Greases that are easily transported to the bearing gap are said to have good 'slumpability'. Such greases require an accurate match between the consistency and the elasticity, quantities which are dependent on the thickener content and the presence of certain polymers.

Figure: Photograph of a test rig with a loaded spiral-groove journal bearing with a transparent Perspex bearing bush. The lubricant (in this case oil) and the boundaries between the lubricant and the air (free boundaries) are clearly distinguishable.

As we have already stated, we prefer an asymmetrical groove pattern for journal bearings so that a reserve of grease can be stored at a predetermined location. This means that there is a free boundary in the bearing gap. The spare grease compensates for any loss of lubricant; this can happen, even with greases, because of effects like creep and evaporation. The life of the grease is also extended if the mechanical and chemical stresses are distributed over a relatively large amount of the grease. One of the requirements for the spare grease is that it can be easily transported to the bearing gap; it must not jell in the reservoir. The chance of this happening is considerable because the shear rate in the reservoir is about 100 times less than in the bearing gap. In other words the grease is 'stirred' much less intensively there. The jelling properties of the grease depend to a large extent on the type and content of thickener used and may also be affected by additives.

While, on the one hand, a grease for use in spiral groove bearings must have good 'slumpability' and must not jell, its yield point must on the other hand always remain high enough to prevent leakage at rest.

This requirement is particularly difficult to meet because the thickener particles in the grease are broken up into even smaller pieces when the bearing is operating. When the bearing stops running, the grease is still warm and thin. It can therefore flow to the unfilled part of the bearing gap because of surface tension ('capillary creep') and it may leak downwards because of gravity. Both types of flow have to be prevented by a sufficiently high yield point. To prevent capillary creep the yield point τ_y must satisfy the condition

$$\tau_y > \gamma/B,$$

where y is the surface tension (about 3 x 10-2 Nyrn) and B the length of the filled part of the bearing. In the vertically mounted spiral-groove journal bearing of Perspex that we used for observing creep and leak age effects this length is 19 mm. To prevent leakage due to gravity it is necessary that:

$$\tau_y > \frac{gQ(\Delta R + h_g)}{2}$$

where g is the acceleration due to gravity (about 10 m/s"), g is the density (about 1 g/cm"), M the clearance (about 30 urn), and hg the groove depth (about 60 urn). Substituting the values for y, B, g, g, M and hg shows that when the yield point decreases during the use of this bearing, capillary creep appears first (at about 1.5 N/m2), and that when the yield point then falls to about 0.5 N/m2, gravity leakage will start.

Figure below shows the qualitative variation in the creep as a function of the time of use for four different greases. In all four it takes some time before the yield point has fallen to such an extent that creep starts. For greases A, C and D the yield point continues to fall and, after a while full creep is attained: the bearing gap is then filledwith grease over its entire length. This is not the case with grease B: after about 1500 hours there is hardly any further increase in the creep. It would seem that the size and the physical behavior of the thickener particles in this grease have by then attained a stable final state. The grease B has a yield point that is ideal for spiral-groove bearings because it ultimately becomes constant at a sufficient magnitude.

If the content of certain polymers in a grease is too high this can lead to an undesirable effect that may increase the chance of leakage considerably: the spare grease may start to 'shuttle'. This effect, which as far as we know has not been observed before, can be clearly seen in an un-loaded horizontally mounted transparent bearing with two groove patterns with about the same pumping action and a grease reservoir on either side. The spare grease, which we shall assume to be in the right-hand reservoir at first, starts to flow towards the left-hand reservoir. This flow continues until the right-hand reservoir is completely empty and a free boundary has been formed in the bearing gap. After a little while the process repeats itself, but in the reverse direction. This can go on for hundreds of hours. We believe that this shuttling movement is due to the 'micro-flow' mentioned earlier and to the effects of the shear rate and time on the viscosity. As a result of the micro-flow 'fresh' grease from the reservoir is exchanged for 'old' grease from the bearing gap. It may be assumed that it will take some time before the viscosity of the fresh grease has adapted to the much higher shear rate (it is about 100 times higher) in the bearing gap. Where delay times are long, this adaptation will be associated with a growing disturbance in the pressure equilibrium in the bearing gap, which makes the grease start to flow in the direction of the empty reservoir.

Changes in Viscosity During The Period of use

When a grease-lubricated spiral-groove bearing is in operation, there is an initial decrease in the viscosity as a result of the breakdown of the thickener. The measured decreases in viscosity can be described as a function of time t, at a constant temperature T, with sufficient accuracy by:

$$\eta(t,T) \approx \eta_0(T)[c_1 + c_2 \exp(-c_3 t)]$$

where 170 is the viscosity of the base oil and Cl, C2and C3 are constants. After a certain time the viscosity assumes an almost constant value because thickener breakdown stops or has no further effect on the viscosity. The motor power, which has to be large enough to overcome the higher initial viscosity, is therefore greater than necessary for the required load carrying capacity and stiffness of the operating bearing. To keep this excess in motor power to a minimum, the decrease in the viscosity must be as small as possible. In figure below the viscosity is shown as a function of the operating time for two of the greases in figure above, A and B. The figure also gives the viscosity of the base oil for each of the two greases.

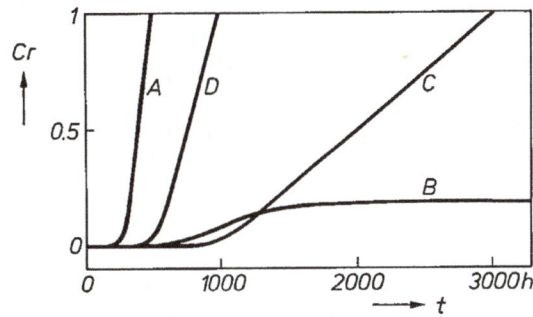

Figure: Qualitative variation of the creep Cr as a function of the test time t for four greases in a spiral-groove journal bearing. Cr = 0: no creep, Cr = 1: full creep, with the longest section of the asymmetrical herringbone pattern also completely filled with grease. Eventually full creep occurs with greases A, C and D, but not with B.

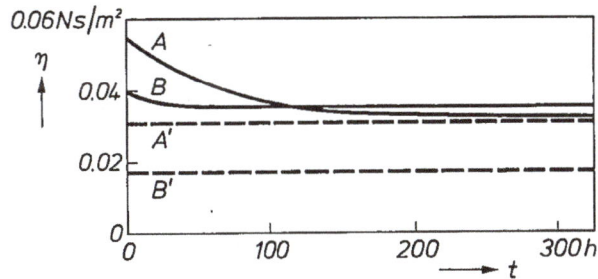

Figure: Viscosity η as a function of the operating time t for greases A and B and their base oils A' and B' at a constant temperature of 55 oe and a constant shear rate of about 5 x 104 S-I. B reveals a much smaller drop in viscosity than A: the viscosity remains considerably higher than that of its base oil B'.

In about 300 hours the viscosity of grease A falls from 0.054 Ns/m2 to a constant value of 0.032 Ns/m", a fall of more than 40070. The viscosity of the grease is then only a few per cent higher than that of its base oil so that there must have been extensive breakdown of the thickener. The viscosity of grease B, on the other hand, decreases within some 50 hours from 0.040 Ns/m" to a constant value of 0.036 Ns/m", a fall of only 10%. In this case the viscosity is still more than twice as high as that of its base oil.

Effect on The Life of Spiral-groove Bearings

The life of a grease-lubricated spiral-groove bearing is limited by a number of factors. The operation of the bearing may start to deteriorate if grease leaks out of the bearing or spare grease jells. Other factors affecting the life are the oxidation of the grease and the evaporation of the base oil. After a time these factors may cause the viscosity to increase considerably, so that the hydrodynamic frictional torque becomes too high. In fig. 16 the frictional torque is shown as a function of operating time for the greases A, B, C and D:GreaseA is known to be very resistant to oxidation. The frictional torque therefore remains constant after more than 10000 hours. With grease B there is a noticeable rise in the torque after about 1000 hours, mainly due to creep and evaporation of the base oil. With grease C oxidation causes an increase in the frictional torque practically from the start. Grease D is very similar in composition to grease C, except that materials have been added to make the grease oxidation-resistant. After about 5000 hours the frictional torque has still not increased.

When a grease that oxidizes easily is used the materials of the shaft and bearing bush can also have a significant effect on the life; Table I illustrates an example of this. This effect is almost always the result of the catalytic effect of copper on the oxidation of greases:-

Table: Example illustrating how the materials of the shaft and bearing bush affect the life of grease-lubricated spiral-groove journal bearings. The results were obtained with a grease that was not very resistant to oxidation. When other greases are used the order may be different and the lives much longer.

Shaft material	Bearing-bush material	Life (hours)
bronze	Perspex	200
Bronze	Steel	250
glass	Perspex	950
glass	Steel	>1000

In finally choosing the grease to be used, a compromise must be found between the minimum creep, the lowest possible fall in viscosity at the start and the smallest possible increase in the frictional torque during long periods of use. Because of its low creep and low fall in viscosity we chose grease B; we are able to reduce the increase in the frictional torque by making a slight change in the composition of the base oil.

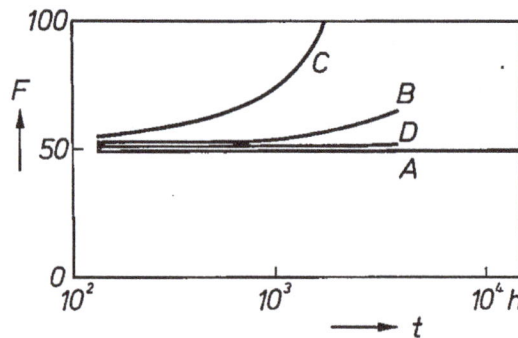

Figure: Frictional torque F (in arbitrary units) as a function of the operating time (of a spiral-groove journal bearing lubricated by one of the greases shown in figure The rise in the frictional torque when B or C is used is the result of an increase in the viséosity. The main reason for this increase for B is loss of base oil and for C it is oxidation.

Manufacture of Spiral-groove Journal Bearings

Bearing Material

Either metal or plastic can be used as the bearing material. For metal spiral-groove bearings we prefer bearing bronze because this is a material that gives good dry-running behavior: if the bearing surfaces come into contact (e.g. on starting or stopping) hardly any damage occurs. The material for plastic bearings must also have good dimensional stability; polyacetals seem to be very suitable.

One of the main disadvantages of plastics is that their thermal conductivity is much worse than that of metals. If the heat generated in the layer of grease is dissipated mainly by way of the bearing bush – and not by way of the metal shaft - the grease may become too hot and hence its viscosity too low, so that the load-carrying capacity and the stability of the bearing may become too small. In addition, the thermal expansion of plastics is so much greater than that of metals that, as the temperature increases, the expansion of the bearing clearance also reduces the load-carrying capacity and the stability. However, good use can be made of this relatively high thermal expansion, for example by enclosing a plastic bearing bush in a metal sleeve; see figure below. Because of the difference in expansion coefficient and modulus of elasticity the bearing gap now becomes smaller instead of larger as the temperature rises, and this compensates for the effects of the fall in viscosity.

When a load is applied, plastics give a greater elastic deformation than metals. Plastics can also undergo plastic deformation as a result of creep effects.

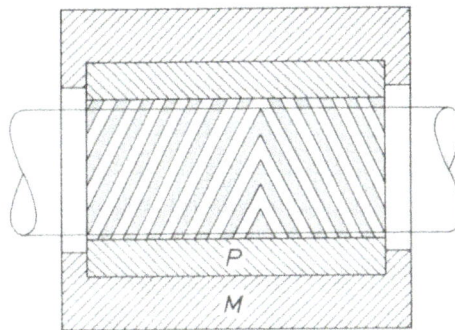

Figure: Diagram of a spiral-groove journal bearing with plastic bearing bush P enclosed in a metal sleeve M. If the plastic wall is thick enough the bearing gap may decrease instead of increase with temperature.

On the other hand, plastics have a greater internal attenuation than metals and therefore transmit vibration and sound less readily between shaft and frame. In equipment in which mechanical vibration must be kept to a minimum (e.g. record players) or where the sound level has to be as low as possible (e.g. electric shavers) this can be an advantage.

The poor electrical conductivity of plastics compared with metals can be both an advantage and a disadvantage. Bearings made of plastic are safer because they are electrically insulating, but metal bearings have the advantage that the suppression of electrical interference is easier.

An advantage of plastics is that they have hardly any effect on the oxidation of greases, so that grease lubricated plastic bearings can have a relatively long life. A practical advantage is that the bearings can be given a color to indicate characteristics that are not immediately obvious such as the groove dimensions or the groove pattern. A practical disadvantage with plastic bearings is that it has not so far been easy to make them with the same dimensional and geometrical accuracy as metal bearings. The result of this is that greater differences in load-carrying capacity and friction will occur in products with plastic bearings for the present than in products with metal bearings. This may mean, for example, that greater attention has to be paid to speed-control systems.

The advantages and disadvantages we have mentioned are so varied that for each particular application it is necessary to consider whether metal or plastic bearings are the more suitable. Nevertheless we do expect that when more experience has been gained with the manufacture and application of spiral-groove bearings, plastic bearings will be increasingly used.

Machining

The two most important machining operations in the manufacture of spiral-groove journal bearings are the production of the bearing bore and the production of the grooves. The bore must be very accurate in dimensions and shape (to within a few urn). The most important requirement for the groove pattern is that the variation in groove depth should be small (also a few urn).

Conventional metal-removal operations can be used for making metal spiral-groove bearings. Both the bearing bore and the grooves can be machined on precision lathes. It is also possible to turn the bearing bore only and to cut the grooves using a tapping process similar to that used for making threaded holes for screws. Particularly accurate bores can be made by broaching. A broach is a tool with an all-round cutting action: a shaft with a 3600 cutting edge is pulled or pushed through a cylindrical hole.

In addition to these metal-removal machining operations there are also forming processes based on plastic deformation. Certain soft and ductile metals can be made to flow around a mandrel by applying a pressure on all sides. Once the pressure has been removed, the formed product springs elastically outwards and can be removed from the mandrel. Plastic deformation may also be imparted by means of rolling balls. Grooves can for example be made in an undersized bore by pressure exerted by rolling balls supported by a cylindrical core. The bore has still to be finish-turned or broached to remove the raised portions.

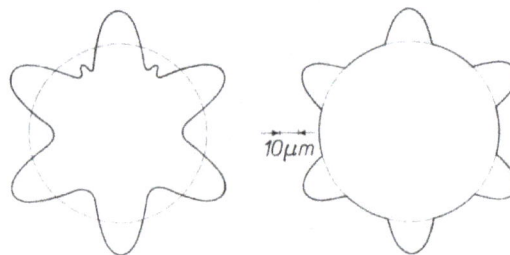

Figure: Results of roundness measurements ('Talyround' measurements) on a bearing bore with rolled-in grooves. Left: not finish turned. Right: finish-turned so that the raised portions have disappeared. The circles denote the original dimension of the bearing bore before the grooves are rolled in.

Figure above shows the results of roundness measurements ('Talyround' measurements) on a bearing bore with rolled-in grooves before and after finish machining.

Physical and chemical metal-removal processes such as spark machining and all kinds of etching operations are less attractive on the whole than mechanical processes for the manufacture of spiral-groove journal bearings.

Injection moulding of thermoplastics is a fast and inexpensive way of producing plastic spiral-groove journal bearings. The injection-moulding process has to be performed under rigidly specified conditions. The temperatures in the mould must be very carefully controlled to obtain the optimum reproducibility. In injection moulding the grooves are made at the same time as the bearing bore. This introduces a complication, since a ridged mould plug has to be removed from a grooved bearing bore.

References

- White, John H. (1985) [1978]. The American Railroad Passenger Car. 2. Baltimore, Maryland: Johns Hopkins University Press. p. 518. ISBN 978-0-8018-2747-1

- Bearing-types, machinery-tools-supplies: thomasnet.co, Retrieved 17 July 2018

- Schwack, F.; Stammler, M.; Poll, G.; Reuter, A. (2016). "Comparison of Life Calculations for Oscillating Bearings Considering Individual Pitch Control in Wind Turbines". Journal of Physics: Conference Series. 753 (11): 112013. doi:10.1088/1742-6596/753/11/112013. ISSN 1742-6596

- Plain-bearings: bearingtips.com, Retrieved 30 June 2018

- "Bicycle History, Chronology of the Growth of Bicycling and the Development of Bicycle Technology by David Mozer". Ibike.org. Retrieved 2013-09-30

- 5-different-types-of-plane-bearings-and-common-uses: tstar.com, Retrieved 20 March 2018

- Brumbach, Michael E.; Clade, Jeffrey A. (2003), Industrial Maintenance, Cengage Learning, pp. 112–113, ISBN 978-0-7668-2695-3

- Types-of-bushings, hardware: thomasnet.com, Retrieved 16 June 2018

- Schwack, Fabian; Byckov, Artjom; Bader, Norbert; Poll, Gerhard. "Time-dependent analyses of wear in oscillating bearing applications (PDF Download Available)". ResearchGate. Retrieved 2017-08-01

- Ball-bearing, technology: britannica.com, Retrieved 11 April 2018

- Roller-bearings: astbearings.com, Retrieved 25 March 2018

Rigging

Rigging refers to the cables, ropes and chains that support a sailing ship or the masts of a sailboat. It is grouped into two categories, standing rigging and running rigging. This chapter discusses in extensive detail about the different types of rigging, such as synthetic, sailboat-running and standing rigging.

Rigging is the term which deals with the sails, masts, booms, yards, stays, and lines of a sailing vessel, or its cordage only.

The basis of all rigging is the mast, which may be composed of one or many pieces of wood or metal. The mast is supported by stays and shrouds that are known as the standing rigging because they are made fast; the shrouds also serve as ladders to permit the crew to climb aloft. The masts and forestays support all the sails. The ropes by which the yards, on square riggers, the booms of fore-and-aft sails, and sails, such as jibs, are manipulated for trimming to the wind and for making or shortening sail are known as the running rigging. The running rigging is subdivided into the lifts, jeers, and halyards (haulyards), by which the sails are raised and lowered, and the tacks and sheets, which hold down the lower corners of the sails. The history of the development of rigging over the centuries is obscure, but the combination of square and fore-and-aft sails in the full-rigged ship created a highly complex, interdependent set of components.

Steam and motor ships commonly carry rigging in the form of masts for supporting hoists, carrying radio antennae, providing lookout mounts, and displaying lights and visual signals.

Synthetic Rigging

Synthetic rigging is disruptive technology that in time will replace stainless steel wire rigging. Since marine surveyors will increasingly come into contact with this type of rigging, they need to understand this new technology to enable them to carry out surveys on craft which use it.

Many new types of synthetic fibres have been discovered in recent years. Typically, they are initially used in aerospace applications and later become available for other application where high performance is required. Most of the high performance fibres are characterized by impressive tensile properties, which with the exception of carbon fibre significantly exceed their compressive strength. With yacht rigs, the mast, spreaders and struts are the only components taking compressive force, and the shrouds (stays) operate as tension only structural members. Therefore, the impressive tensile properties of these fibres make them ideal for standing rigging.

There are always new, stronger fibres being developed, however currently available fibres used for synthetic rigging include:

- Synthetic (PBO)

- Synthetic (Carbon Fibre)

- Synthetic (Aramid)

- Synthetic (HMPE)

Synthetic rigging will replace wire and marine surveyors need to understand this disruptive new technology

PBO (Zylon): PBO is short for (polybenzoxazole) is a trademarked name for a range of thermoset liquid-crystalline polyoxazole. This synthetic polymer material was invented and developed by SRI International in the 1980s and is manufactured by the Toyobo Corporation. In generic usage, the fibre is referred to as PBO.

Carbon Fibre: Carbon Fibres are long parallel chains of carbon atoms that are formed by stretching and heating certain forms of organic filaments. Carbon fibre laminates have fatigue limits far in excess of steel and excellent vibration damping characteristics, but have poor impact strength. Carbon fibre is commonly used high performance fibre which is extensively used in the marine industry for high performance structures including hulls, masts and rigging.

Aramid: A para-aramid synthetic fiber with the trade name Kevlar was developed at DuPont in 1965 and first used commercially in the early 1970s. A similar fiber to Kevlar called Twaron with roughly the same chemical structure was developed by Akzo in the 1970s with commercial production started in 1986. Twaron is now manufactured by Teijin.

HPME: Spectra fibres were first introduced into the marketplace in 1985, after a decade of intensive research, engineering and development by the Allied Fibres division of Allied Signal Technologies. Spun from a solution of Ultra High Molecular Weight Polyethylene (UHMWPE), HPME fibres combine a very high degree of molecular orientation with a very low density which results in fibres with unique and quite extraordinary performance profile. Even among the so-called high performance fibres, the unique physical properties of HPME place it in a class of its own. It is marketed under the trade names of Spectra and Dyneema.

Most forms of synthetic rigging exhibit stretch over time (creep). However, creep can and is controlled by appropriate design with stretch equivalent sizing being used.

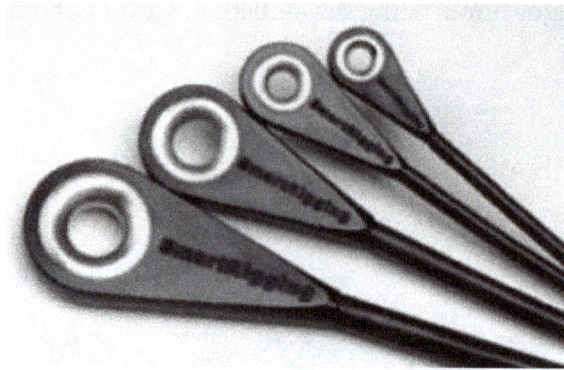

Figure: Factory sheathed and terminated parallel aramid rigging. Picture courtesy Aramid Rigging

Different approaches are used to create synthetic rigging solutions:

- Continuous rigging: Using continuous fibres the vertical and diagonal shroud elements are fused to form a single, homogeneous piece of rigging.

- Endless winding of single elements: Using a process that involves continuous winding of fibres around two thimbles until the target cable strength or required cable stretch has been reached.

- Fibre rod rigging: Using thin pultruded rods, bundled together to achieve target strength.

- Fibre solid rod rigging: Using rigging elements that are formed as a solid rod like traditional rod rigging, not bundled.

- Rope rigging:

  ○ Fibres in parallel strand form encased in a protective polymeric sheath.

  ○ Fibres in a rope form typically 12 strand.

With continuous rigging and endless winding, fibres are wound around terminals as part of the manufacturing process. When using fibre rod rigging the fibre bundle is typically bonded into terminal fittings. While with rope rigging terminals of similar concept to Norseman Sta-Lock are used for parallel fibre type while the stranded rope form is spliced around special thimbles similar in concept to Merriman thimbles.

Figure: Special thimbles used with spliced termination on stranded HPME rope rigging. Pictures courtesy Colligo Marine

With the exception of Carbon Fibre solid rod rigging and HPME stranded rope rigging, the fibers are packaged in some form of sheathing to protect the product from physical damage and exposure to the elements, particularly moisture and UV light (figure 4). Stranded rope rigging made from HPME fibers is not adversely affected by moisture and UV light so it is not sheathed, but may be surface treated.

Synthetic rigging is gaining acceptance and widespread use on sailing craft of all forms, from monohull to multihull, racer and cruiser with typical usage being indicated in table.

The advantages of synthetic fibre rigging are:

- significant gains in strength

- significantly lower weight aloft

- ability to easily inspect (when using synthetic rope rigging)

PBO rigging has had over a decade of proven success at the America's Cup level. As an example, PBO is claimed to be 65% lighter than traditional rigging at 110–130% of the price of rod rigging.

Figure: Sheathed Aramid Rigging. Picture courtesy Aramid Rigging

Table: Typical Usage

Type of Rigging	Usage			
	Racer	Racer/Cruiser	Cruiser/Racer	Cruiser
Synthetic(PBO)	✔	✔		
Synthetic (Carbon Fibre)	✔ (high/intermediate modulus)	✔ (standard modulus)		
Synthetic(Aramid)	✔	✔	✔	✔
Synthetic(HMPE)	✔	✔	✔	✔

PBO, Carbon and Aramid rigging as it is made to specific length, terminated and sheathed using specific processes and equipment, it is typically provided by specialist rigging companies as a manufacture and fit service.

HMPE particularly that based on stranded Dyneema SK75 rope rigging is cost effective for cruiser and cruiser/racer craft. HMPE specifically Dyneema SK75 is typically sold under trade names Dynice DUX (Hampidjan) and STS-12 (New England Ropes) and is braided rope of 12-strand construction. Dynice DUX is heat set which removes constructional elongation and further reduces stretch increasing its performance and suitability for use as standing rigging. Since the ends are spliced and since companies like Colligo Marine in the USA manufacture and supply innovative connection hardware it has allowed this HMPE rigging to be provided by smaller rigging companies or done by individuals as a DIY project.

Figure: HPME stranded rope rigging (Dynice DUX) spliced around special thimbles and attached to (turnbuckles) rigging screws. It should be noted that the shroud in the foreground (without black chafe protection can easily be fully inspected. Picture courtesy of Colligo Marine

Synthetic rigging results in a more reliable rig, a stiffer, faster boat with greatly reduced pitching moment. Reducing weight aloft on any sailing craft lowers the vertical position of the center of gravity (VCG), which dramatically improves a yachts handling, stability, and response. Consequently, moving to synthetic rigging can represent a cost effective "upgrade" for any type of sailing craft. Based on acceptance and use of affordable synthetic rigging marine surveyors will increasingly surveying craft using HMPE rigging. Inspection of HMPE rigging is easier and outcomes are more conclusive than traditional stainless steel wire rope rigging inspections as all aspects of terminations are visible.

Lifespan of synthetic rigging, if well cared for, is expected to be at least as good as wire rigging, however many rigging suppliers based on specific experience indicate longer lifespans. Since the use of synthetic rigging is relatively new, data regarding lifespan is still being gathered.

Sailboat-Running Rigging

Sailboat rigging can be described as being either running rigging which is adjustable and controls the sails - or standing rigging, which fixed and is there to support the mast. And there's a huge amount of it on the average cruising boat.

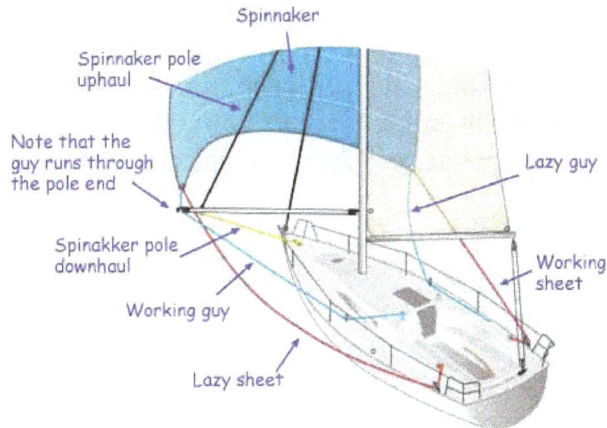

You'll need a whole lot more of it if you fly a spinnaker.

- Port and starboard sheets for the jib, plus two more for the staysail (in the case of a cutter rig) plus a halyard for each - that's 6 separate lines;

- In the case of a cutter you'll need port and starboard runners - that's 2 more;

- A jib furling line - 1 more;

- An up-haul, down-haul and a guy for the whisker pole - 3 more;

- A tackline, sheet and halyard for the cruising chute if you have one - another 3;

- A mainsheet, halyard, kicker, clew outhaul, topping lift and probably three reefing pennants for the mainsail (unless you have an in-mast or in-boom furling system) - 8 more.

The Essential Properties of Lines for Running Rigging

It's often under high load, so it needs to have a high tensile strength and minimal stretch.

It will run around blocks, be secured in jammers and self-tailing winches and be wrapped around cleats, so good chafe resistance is essential.

Finally it needs to be kind to the hands so a soft pliable line will be much more pleasant to use than a hard rough one.

Not all running rigging is highly stressed of course; lines for headsail roller reefing and mainsail furling systems are comparatively lightly loaded, as are mainsail jiffy reefing pennants, single-line reefing systems and lazy jacks.

But a fully cranked-up sail puts its halyard under enormous load. Any stretch in the halyard would allow the sail to sag and loose its shape.

It used to be that wire halyards with spliced-on rope tails to ease handling were the only way of providing the necessary stress/strain properties for halyards.

Thankfully those days are astern of us - running rigging has moved on a great deal in recent years, as have the winches, jammers and other hardware associated with it.

Modern Materials

Ropes made from modern hi-tech fibres such as Spectra or Dyneema are as strong as wire, lighter than polyester ropes and are virtually stretch free. It's only the core that is made from the hi-tech material; the outer covering is abrasion and UV resistant braided polyester.

But there are a few issues with them:

- They don't like being bent through a tight radius. A bowline or any other knot will reduce their strength significantly.

- For the same reason, sheaves must have a diameter of at least eight times the diameter of the line.

- Splicing securely to shackles or other rigging hardware is difficult to achieve, as it's slippery stuff. Best to get these done by a professional rigger.

Approximate Line Diameters for Running Rigging

But note the word 'approximate'. More precise diameters can only be determined when additional data regarding line material, sail areas, boat type and safety factors are taken into consideration.

Length of boat	6m (20ft)	8m (26ft)	10m (33ft)	12m (40ft)
Spinnaker guys	10mm	10 to 12mm	12 to 14mm	16mm
Boom Vang and pre-venters	8mm	8mm	10mm	12mm
Spinnaker sheet	6 to 8mm	6 to 10mm	8 to 12mm	10 to 14mm
Main sheet	8 to 10mm	10mm	10 to 12mm	12 to 14mm
Genoa sheet	10 to 12mm	12mm	14mm	14mm
Jib sheet	10mm	10 to 12mm	12mm	14mm
Main halyard	6 to 8mm	8mm	10mm	12mm
Genoa / Jib halyard	6 to 8mm	8mm	10mm	12mm
Spinnaker halyard	6 to 8mm	8mm	10mm	12mm
Pole uphaul	8mm	8mm	8mm	10mm
Pole downhaul	8mm	8mm	10mm	12mm
Reefing pennants	8mm	8mm	10mm	12mm

Lengthwise it will of course depend on the layout of the boat, the height of the mast and whether it's a fractional or masthead rig - and if you want to bring everything back to the cockpit.

Standing Rigging

Standing rigging is the permanent rigging (such as stays and shrouds) used primarily to secure the masts and fixed spars of a vessel or to support radio, radar, and other equipment carried aloft.

The Standing Rigging consists of the wires that hold up and support the mast. Because the mast is in compression and tends to buckle or bend, the standing rigging helps to control the bending. Some small sailboats do not use any standing rigging, and these are said to have free standing un-stayed masts. The calculations and methods of figuring the strength of spars and associated rigging are very technical and involved, and should not be undertaken by the novice.

The material used for the standing rigging is wire rope, usually made from stainless steel, although regular or galvanized steel wire rope is available. Wire rope is measured by the diameter and spec-ified by the composition of the wires used to make up the wire rope. For example, wire rope des-ignated 1 x 19 would consist of one wire made up of 19 strands. This type is the most common for standing rigging because it is not flexible and is strongest. Another type designated 7 x 19 consists of 7 ropes each consisting of 19 strands. This type, while not as strong, is used where flexibility is important. On boats which use wire rope halyards, the 7 x 19 wire rope is utilized.

Stays

The STAYS are wire ropes that support the mast in a fore and aft direction. Technically speaking, any wire that helps support the mast can be called a stay. The Forestay supports the mast from the forward side and is usually attached to the hull near the forward end of the boat. On a jibhead rig the forestay is attached to the mast about 7/8 the way up from the base. On a masthead rig the forestay attaches to the masthead. On catamarans because of the twin hulls, the forestay often in-tersects with a Bridle, and the bridle is attached to the bow of each hull. A bridle is a line secured at each end with attachment by another line to the middle of the bridle. Some catamarans use a beam between the hulls at the bow and attach the forestay to this beam at the middle in the conventional manner. On single hull boats, the forestay is connected to the hull at the Stemhead (forward point of the hull usually at the deck). The forestay must be capable of withstanding considerable strain. The other stay on masthead rigs that complements the forestay is the Backstay. The backstay sup-ports the mast from the aft side, and runs from the masthead to the aft end of the boat.

The stays that support the mast at the sides are called Shrouds, and it is not correct to call them "side stays" or any other name when they are being referred to specifically. As mentioned previous-ly, not all boats use stays, but boats using a forestay will invariably have at least one shroud per side.

When the mast requires additional support, two or more sets of spreaders are required, especially on boats which also use a backstay. The set of shrouds which pass through or across the spreader tips and attach to the masthead are called the upper shrouds. They are usually located in line with the side of the mast functioning along the gunwale or rail of the hull or cabin side. The shrouds that join to the mast at the spreader connection are called the lower shrouds, and may connect to the hull forward or aft of the upper shrouds. In some boats it is not uncommon to use two sets of lower shrouds, joining the hull at least several inches apart from each other outboard. The reason shrouds should preferably not junction at a common point is in order to distribute the mast loads over a greater area of the hull.

The shrouds usually attach to the hull via the chain plates, while the forestay attaches to the stem head fitting, and the backstay to a backstay tang or chain plate. At the spreaders, the upper shrouds should be protected from chafing where they move at the spreader tips. This is best accomplished by using non-chafing spreader tips. The spreader tips themselves should be rounded or smoothed so as not to chafe or snag the sails.

Fasteners

A fastener is a hardware device, which affixes or joins objects together. Fasteners create non-permanent joints. Some of the varied fasteners used in machinery are mechanical fasteners, threaded fasteners and screws, which have been discussed in depth in this chapter.

A fastener is a hardware device that mechanically joins or affixes two or more objects together. It is defined as a hardware which can be easily installed and removed with hand tool or power tool. Common fasteners include screws, bolts, nuts and rivets. The terms bolts and screws do not refer to specific types of fasteners, but rather how they are used (i.e. the application). Thus the same fastener may be termed a bolt or a screw. Bolts are defined as headed fasteners having external threads that meet an exacting, uniform thread specification such that they can accept a non-tapered nut. Screws are defined as headed, externally-threaded fasteners that do not mate with a non-tapered nut and are instead threaded into the material they hold.

Principal purposes of the fasteners are (i) dis-assembly for inspection and repair, (ii) modular design, where a product consists of a number of sub-assemblies.

The fastener types are (i) removable which permits the parts to be readily disconnected without damaging the fastener, e.g. nut and bolt, (ii) semi-permanent type where the parts can be disconnected, but some damage usually occurs to the fastener, e.g. cotter pin, and (iii) permanent type where the parts are never be disassembled e.g. rivets and welding of fasteners

The most common types of male fasteners used in industry are hex head, slotted head, flat (or countersunk) head, round head, socket (or 'allen') head, button head and socket set screw. The most common types of female fasteners (i.e. nuts) used in industry are regular hexagonal nuts and nylon ring elastic stop nuts (also known as 'lock nuts').

Fasteners have only one intended function which is to clamp two parts together. Fasteners are not meant to position parts relative to one another. They are also not meant to function as pivots, axles and fulcrums.

More importantly, the threaded portion of a fastener is not to be loaded in shear for at least three reasons namely (i) the threaded portion of the fastener is of slightly smaller diameter than the unthreaded shank, allowing the fastener to quickly loosen if transverse loading is applied, (ii) the threaded portion of the bolt has much less surface area than the shank which means it offers significantly less bearing area to the joint and this reduces the load carrying capacity and fatigue resistance of the assembly, and (iii) when the relative motion between the hole and the loose fitting threaded portion of the bolt occurs, the thread acts as a low speed file, removing material from the inside of the hole, aggravating the problem.

There is no fastener material that is right for every environment. Selecting the right fastener material from the vast array of materials available appears to be a daunting task. Careful consideration may need to be given to strength, temperature, corrosion, vibration, fatigue and many other variables.

Mechanical Properties

Most fastener applications are designed to support or transmit some form of externally applied load. If the strength of the fastener is the only concern, there is usually no need to look beyond carbon steel.

The most widely mechanical property associated with standard threaded fasteners is tensile strength. Tensile strength is the maximum tension-applied load which the fastener can support prior to or coinciding with its fracture.

The proof load represents the usable strength range for certain standard fasteners. By definition, the proof load is an applied tensile load that the fastener must support without permanent deformation. In other words, the bolt returns to its original shape once the load is removed. The steel possesses a certain amount of elasticity as it is stretched. If the load is removed and the fastener is still within the elastic range, the fastener always returns to its original shape. If, however, the load applied causes the fastener to be brought past its yield point, it now enters the plastic range. Here, the steel is no longer able to return to its original shape if the load is removed. The yield strength is the point at which permanent elongation occurs. If the load is continued to be applied to reach a point of maximum stress (ultimate tensile strength) then after this point is reached the fastener necks and elongates.

Shear strength is defined as the maximum load that can be supported prior to fracture, when applied at a right angle to the axis of the fastener. A load occurring in one transverse plane is known as single shear. Double shear is a load applied in two planes where the fastener can be cut into three pieces. For most standard threaded fasteners, shear strength is not a specification even though the fastener may be commonly used in shear applications. While shear testing of blind rivets is a well-standardized procedure which calls for a single shear test fixture, the testing technique of threaded fasteners is not as well designed. Most procedures use a double shear fixture, but variations in the test fixture designs cause a wide scatter in measured shear strengths. There are two possibilities for the applied shear load. One has the shear plane corresponding with the threaded portion of the bolt. Since shear strength is directly related to the net sectional area, a smaller area results in lower bolt shear strength. To take full advantage of strength properties, the preferred design is always to position the full shank body in the shear planes.

A fastener subjected to repeated cyclic loads can suddenly and unexpectedly break, even if the loads are well below the strength of the material. The fastener fails in fatigue. The fatigue strength is the maximum stress a fastener can withstand for a specified number of repeated cycles prior to its failure.

Torsional strength is a load usually expressed in terms of torque, at which the fastener fails by being twisted off about its axis. Tapping screws and socket set screws are usually tested for torsional strength.

Hardness is a measure of a material's ability to resist abrasion and indentation. For carbon steels,

Brinell and Rockwell hardness testing is generally used to estimate tensile strength properties of the fastener.

Ductility is a measure of the degree of plastic deformation that has been sustained at fracture. In other words, it is the ability of a material to deform before it fractures. A material that experiences very little or no plastic deformation upon fracture is considered brittle. A reasonable indication of a fastener's ductility is the ratio of its specified minimum yield strength to the minimum tensile strength. The lower this ratio the more ductile the fastener is.

Toughness is defined as the ability of a material to absorb impact or shock loading. Impact strength toughness is rarely a specification requirement. Besides various industry fasteners need low temperature service. These fasteners require low temperature impact testing.

Materials for Fasteners

Over 90 % of all fasteners are made of carbon steel. In general, considering the cost of raw materials, nonferrous material for the fasteners is generally considered only when a special application is needed.

Carbon steel has excellent workability, offers a broad range of attainable combinations of strength properties, and in comparison with other commonly used fastener materials, is less expensive. The mechanical properties are sensitive to the carbon content. For fasteners, the more common steels are generally classified into three groups. These are namely (i) mild steel, (ii) medium carbon and (iii) alloy steel.

Mild steels generally contain less than 0.25 % carbon and cannot be strengthened by heat-treatment. Strengthening may only be accomplished through cold working. The mild steel material is relatively soft and weak, but has outstanding ductility and toughness; in addition, it is machinable, weldable and is relatively inexpensive to produce.

Medium carbon steels have carbon concentrations in the range of 0.25 % to 0.60 %. These steels can be heat treated by austenizing, quenching and then tempering to improve the mechanical properties. The plain medium carbon steels have low hardenabilities and can be successfully heat-treated only in thin sections and with rapid quenching rates. This means that the end properties of the fastener are subject to size effect. On a strength-to-cost basis, the heat-treated medium carbon steels provide tremendous load carrying ability. They also possess an extremely low yield to tensile strength ratio; making them very ductile.

Carbon steel can be classified as alloy steel when the manganese content exceeds 1.65 %, when silicon or copper exceeds 0.60 % or when chromium is less than 4 %. Carbon steel can also be classified as an alloy steel if a specified minimum content of aluminum, titanium, vanadium, nickel or any other element has been added to achieve specific results. Additions of chromium, nickel and molybdenum improve the capacity of the alloys to be heat treated, giving rise to a wide variety of strength to ductility combinations.

Stainless steel is a family of iron-based alloys that must contain at least 10.5 % chromium. The presence of chromium creates an invisible surface film that resists oxidation and makes the material 'passive' or corrosion resistant. Other elements, such as nickel or molybdenum are added to increase corrosion resistance, strength or heat resistance. Stainless steels can be simply and logically divided into three classes on the basis of their microstructure namely (i) austenitic, (ii) martensitic, or (iii) ferritic. Each of these classes has specific properties and different grades.

Also, further alloy modifications can be made to alter the chemical composition to meet the needs of different corrosion conditions, temperature ranges, strength requirements, or to improve weldability, machinability, work hardening and formability.

The family of nickel alloys offer some remarkable combinations of performance capabilities. Mechanically they have good strength properties, exceptional toughness and ductility, and are generally immune to stress corrosion. Their corrosion resistance properties and performance characteristics in both elevated and sub-zero temperatures is superior. Unfortunately, nickel based alloys are relatively expensive. The two most popular nickel alloys used in fastening are the nickel-copper and nickel-copper-aluminum types.

Inconel and Hastelloy are considered outstanding materials for applications where fastenings must contain high strength and resistance to oxidation in extreme environments such as elevated temperatures and various acidic environments. There are several grades of Inconel and Hastelloy, most are proprietary, and practically all are trade named, each with their own strength and corrosion characteristics.

Aluminum is synonymous with lightweight. Once thought as only a single costly metal, aluminum now constitutes an entire family of alloys. Aluminum can be alloyed with other metals to produce suitable alloys with variety of industrial and consumer goods. Aluminum fasteners weigh about one third those of steel.

Silicon bronze is the generic term for various types of copper-silicon alloys. Most are basically the same with high percentages of copper and a small amount of silicon. Manganese or aluminum is added for strength. Lead is also added for free machining qualities where required. Silicon bronze possesses high tensile strength (superior to mild steel). With its high corrosive resistance and non-magnetic properties, this alloy is ideally suited for naval construction particularly in mine sweepers.

Naval bronze, sometimes called naval brass, is similar to brass but has additional qualities of resistance to saline elements. This is accomplished by changing the proportions of copper, zinc and a little tin. This alloy derived its name from its ability to survive the corroding action of salt water.

Copper has some very interesting performance features. Its electrical and thermal conductivity are the best of any of non-precious metals and has decent corrosion resistance in most environments. Copper, and its alloys, are non-magnetic.

Brass is composed of copper and zinc and is the most common copper-based alloy. They retain most of the favourable characteristics of pure copper, with some new ones, and generally cost less. The amount of copper content is important. Brass alloys with less copper are generally stronger and harder, but less ductile.

Fasteners Nomenclature

Design engineers are frequently required to select and specify fasteners used in their designs.

Consequently, understanding basic fastener nomenclature is important. Fig 1 illustrates the different parts of a standard threaded fastener.

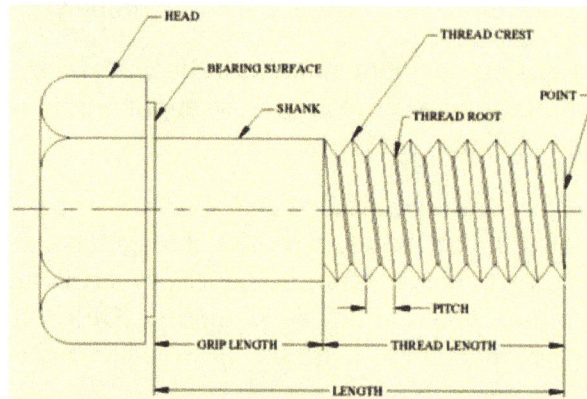

Fig: Important nomenclature of a male fastener

The fastener nomenclature terminology is described below.

- Major diameter: It is the largest diameter of a fastener thread;

- Minor diameter: It is the smallest diameter of a fastener thread;

- Pitch: It is the linear distance from a point on the thread to a corresponding point on the next thread: measured parallel to the axis of the thread;

- Lead: It is the linear distance that a point on a fastener thread advances axially in one revolution (equal to the pitch of the fastener);

- Thread root: It is the surface of the thread that joins the flanks of adjacent threads and is immediately adjacent to the cylinder from which the thread projects. In other words, it is the valley of the thread;

- Thread crest: It is the surface of the thread that joins the flanks of the thread and is farthest from the cylinder from which the thread projects. In other words, it is the peak of the thread;

- Head: It is the enlarged shape that is formed on one end of the fastener to provide a bearing surface and a method of turning (or holding) the fastener;

- Bearing surface: It is the supporting surface of a fastener with respect to the part it fastens;

- Point: It is the extreme end of the threaded portion of a fastener;

- Shank: It is the cylindrical part of a fastener that extends from the underside of the head to the starting thread;

- Length: It is the axial distance between the bearing surface of the head and the extreme point;

- Grip length: It is the length of the unthreaded portion of the fastener (i.e. shank) measured axially from the underside of the bearing surface to the starting thread;

- • Thread length: It is the length of the threaded portion of the fastener. In all the commercial fasteners, threaded length is a function of fastener diameter.

Fastener Thread Types

In the most general sense, there are two classes of fastener threads namely (i) English, and (ii) metric. For each class there are two types of threads namely (i) fine thread, and (ii) coarse thread.

One of the most common fastener mistakes is using the wrong type of thread in a material. The basic rule for fastener selection is namely (i) fine threads are stronger when the female thread is strong relative to the male thread, and (ii) coarse threads are stronger when the female thread is weak relative to the male thread. The reason for this statement is that a smaller minor diameter increases the thread area, resulting in higher static strength and fatigue resistance in female threads. Conversely, a larger minor diameter increases the stress area, resulting in a higher static strength and fatigue resistance in male threads. When fasteners are selected then it is assumed that the stresses are distributed over only the first five engaged threads.

Due to the elasticity of the fastener, only the first five threads are engaged during loading regardless of the thread type (coarse / fine). Female threads typically fail due to shear along the major diameter and male threads typically fail due to tensile loading along the thread root.

Since five threads carry the entire load regardless of thread type, a decrease in the minor diameter increases the shear area and gives an advantage to the female threads while reducing the load carrying capability of the male fastener. Conversely, an increase in the minor diameter increases the cross sectional area of the male fastener and gives an advantage to the male fastener. However, this reduces the shear area and weakens the female threads. Therefore, if the female fastener material is weak compared to the male fastener material, the female fastener is to be given the advantage and coarse threads are to be chosen. If the female fastener material is strong compared to the male fastener material then the male fastener always fails first and hence is to be given the advantage by selecting fine threads.

For this reason steel bolts and studs that thread into relatively weak aluminum or cast iron castings such as engine blocks, cylinder heads and gearboxes are always coarse threaded on the end that goes into the casting. Also invariably, the end of the stud that receives the nut is provided with a fine thread. In this way the designer ends up with the best of both worlds.

Because coarse threads are faster to assemble, they are often used in applications where strength and weight are not of utmost concern. Generally, unless threading into a relatively weak material, the coarse threaded fasteners are to be avoided.

Comparison of Rolled Threads and Cut Threads

All quality fasteners have rolled threads produced via rolling or sliding dies. Rolled threads (as opposed to threads cut on a lathe, with a cutting die or tap) produce superior surface finish (thus lower stress risers) and improved material properties from cold working the material, resulting in much higher fatigue resistance. Rolled threads increase thread strength by a minimum of 30 % over well-cut threads.

When a thread is cut into a specimen, the grain flow of the material is severed. When a thread is rolled into a specimen, however, the grain flow of the material remains continuous and follows the contour of the thread. For this reason, rolled threads better resist stripping because shear failures must take place across the material grain rather than with it. Another benefit of thread rolling is it produces a much better surface finish than thread cutting. In high strength fasteners, rolled threads possess up to twice the fatigue resistance compared to cut threads.

Rolling also leaves the surface of the threads, particularly in the roots, stressed in compression. These compressive stresses must be overcome before the tensile stresses can reach a level that will cause fatigue failures. Compressive surface stresses also increase root hardness, further adding to the fatigue resistance of the component. Improved fatigue strength resulting from the above factors is reported to be on the order of 50 % – 75 %. On heat-treated bolts that have threads rolled after heat-treatment, tests show increased fatigue strength of 5 to 10 times that of cut threads.

All quality fasteners must be heat-treated to achieve the desired strength and toughness. The heat-treatment process inevitably results in some physical distortion of the fastener blank. Rolling the thread onto the (already) heat treated blank ensures the thread will be coaxial with the bolt and normal to the bearing surface of the fastener head, which is critical for proper function. Finally, due to the speed at which fastener threads can be rolled onto a blank with the proper equipment, rolled threads can actually be more economical to manufacture in larger quantities.

Platings and Coatings

Most of the threaded fasteners used today are coated with some kind of material as a final step in the manufacturing process. Many are electroplated, others are hot-dipped or mechanical galvanized, painted or furnished with some other type of supplementary finish. Fasteners are coated for four primary reasons namely (i) for appearance, (ii) to fight corrosion, (iii) to reduce friction, (iv) to reduce scatter in the amount of preload achieved for a given torque.

There are three basic ways in which coatings can fight corrosion. These are follows:

- They can provide a barrier protection. This simply means that they erect a barrier, which isolates the fastener from the corrosive environment, thereby breaking the metallic circuit, which connects the anode to the cathode.

- They can provide a 'galvanic' or sacrificial protection. To cause problems, a metallic connection must be made between the anode and the cathode and an electrolyte. In this type of reaction it is always the anode, which will get attacked, so if the fastener is protected by making it the cathode.

- They can fight corrosion by 'passivation' or 'inhibition', which slows down the corrosion and makes the battery connection less effective. This is common with the use of nickel in stainless steel bolts, which are said to be passivated. A thin oxide layer is formed on the surface of the bolt. The oxide film, according to theory, makes it more difficult for the metal to give off electrons.

Mechanical Fasteners

Most of the threaded fasteners used today are coated with some kind of material as a final step in the manufacturing process. Many are electroplated, others are hot-dipped or mechanical galvanized, painted or furnished with some other type of supplementary finish. Fasteners are coated for four primary reasons namely (i) for appearance, (ii) to fight corrosion, (iii) to reduce friction, (iv) to reduce scatter in the amount of preload achieved for.

Types of Mechanical Fasteners in Agriculture

Rivets are known as permanent mechanical fasteners, and there are a number of types of rivets that a worker can choose from. They are installed through a hole that is drilled into the surface that this tool is put through. After the head is placed into the hole, the tail will expand to ensure that the rivet doesn't move and that the material it is attached to stays in place.

The J-Bolt Has Many Different Functions

If you need to build a barn, silo or other structure that is built to last, you will want to use a J-bolt. It is among the strongest of the different types of mechanical fasteners on the market. The J-bolt is often used to hang electrical wire or sheet metal, and it can be used in concrete and other heavy materials. The bolt itself can also come in a variety of different materials to meet your needs.

Indented Hex Head Screws Make Fastening Easier

Among the permanent fasteners types that a person could use, the indented hex head screws are the most convenient. Compared to other fasteners, this product creates more natural torque, which means that they can be installed without as much physical force being applied by an individual. In some cases, there is no need to use a nut or other object to secure them in place.

Pivot Pins Are for Semi-Permanent Structures

When you need a temporary pen for animals that are giving birth or a similar structure, they can easily be built with pivot pins. The pin itself has a plunger on top that can be taken off. From there, the pin can be easily removed with little force exerted. This can be ideal if you discover that there are gaps within the structure or other issues that need to be fixed.

These Tools Can be used in a Variety of Climates

One of the benefits of using these fasteners is that they can be useful in a variety of different climates. This is ideal for those who live in parts of the country that experience all four seasons throughout the year or experience temperature swings over the course of several days. Since these fasteners can withstand the conditions, it means that farmers and other workers spend less time on maintenance work.

Having an understanding of the different types of fasteners and their uses can save a person time and money. By letting the fasteners do the work of connecting and holding heavy structures together, there is less of a chance that an individual will get hurt while working. It also allows individuals to focus more on their core job as opposed to miscellaneous building tasks.

Threaded Fasteners

Threaded fasteners are commonly used for a multitude of reasons. Aside from the obvious strength benefits (a fastener properly threaded or a bolted joint is much stronger than unthreaded alternatives) they are also reversible, meaning they can be taken apart after they are assembled. They are cost-efficient and available in a wide range of sizes.

Because threaded fasteners like bolts, nuts, and washers can be used many times in a single application, they can add extra weight to a product. Additionally, they often need retightening when located in awkward places. They are susceptible to vibrations, and may need safety wire or other added features to further hold them in place and enable proper performance.

There is a wide array of threaded fastener types, including:

- Unified national coarse (UNC)

- Unified national fine (UNF)

- Unified national extra fine (UNEF)

- UNJC and UNJF threads

- UNR threads

- UNK threads

- Constant-pitch threads

Unified National Coarse Threads

UNC threads are the most common general fastener thread. Their fit is deeper and more generic than that of a fine thread, allowing for easy removal. Generally, they have a higher tolerance for manufacturing and plating, and do not need cross threading to assemble.

Unified National Fine Threads

UNF threads have better torque-locking and load-carrying ability than UNC threads because of their larger minor diameter. Because of their more specific fit, they have tighter tolerances, finer tension adjustment, and can carry heavier loads. They are most commonly found in the aerospace industry.

United National Extra Fine Threads

UNEF threads are finer than UNF threads; they are used in applications with tapped holes in hard material, thin threaded walls, and tapped holes in thin material. As with UNF threads, UNEF threads are common in the aerospace industry.

UNJC and UNJF Threads

There are two types of "J" threads: external and internal. External UNJC and UNJF threads have a larger root radius than the corresponding part (either UNC, UNR, UNK, or UNF threads). The larger root radius results in a larger tensile area than the corresponding thread, and smaller stress concentration—bolts that carry heavy loads usually use "J" threads.

UNR and UNK Threads

A UNR external thread is the same as a UNC thread, only the root radius is rounded. There is no internal UNR thread. UNK threads resemble UNR threads, but the root radius and minor diameter require inspection.

Constant-Pitch Threads

These threads come in a variety of diameters to fit a given application—bolts with diameters of 1 in. and above commonly use pitches of 8, 12, or 16 threads per inch.

Internal threads refer to those on nuts and tapped holes, while external threads are those on bolts, studs, or screws.

Threads can be produced by either cutting or rolling.

Cut threading is a process by which steel is cut away from a round bar of steel to form the threads.

Advantages of Cut Threading:

1. Few limitations with regard to diameter and thread length.

2. All specifications can be manufactured with cut threads.

Disadvantages

1. Longer labor times.

2. The weakest area of any mechanical fastener is the minor diameter of the threads. Since the thread dimensions of a cut thread and rolled thread fastener are identical, there is absolutely no difference in strength. However, cut threading interrupts the natural grain structure of the round bar whereas roll threading reforms it.

Advantages of Roll Threading

1. Shorter labor times.

2. Weighs less than its Cut threaded bolts thereby reducing the cost of the steel, plating, freight, and any other costs associated with the fastener that are based on weight.

3. Cold working makes threads more resistant to damage during handling.

4. Rolled threads are often smoother. The distance measured parallel to the thread axis, between corresponding points on adjacent threads, is the thread pitch. Unified screw threads are designated in threads per inch. This is the number of complete threads occurring in one inch of threaded length. Metric thread pitch is designated as the distance between threads (pitch) in millimeters.

A screw/bolt is a type of fastener distinguished by a circular ridge, known as threads which are wrapped around a cylinder.

While some screw threads are designed to get along with an internal thread, also known as complementary thread, there are some threads, which are fashioned to cut a helical groove in a softer material, say wood or plywood, when the fastener is inserted. The major function of a screw is to hold two objects together.

These kinds of fasteners are astonishingly versatile and powerful enough to be hold two different surfaces intact. The basic concept behind such metal objects' functionality is their threaded cylindrical parts, which are designed to get inserted into any kind of material, be it a plastic, wood, metal or plywood. However, there are various types of screws available suitable for distinct surfaces.

These types can be distinguished depending upon the driving methods, job requirement, head shape, type of threads and material used to make such metal pieces.

Driving Methods

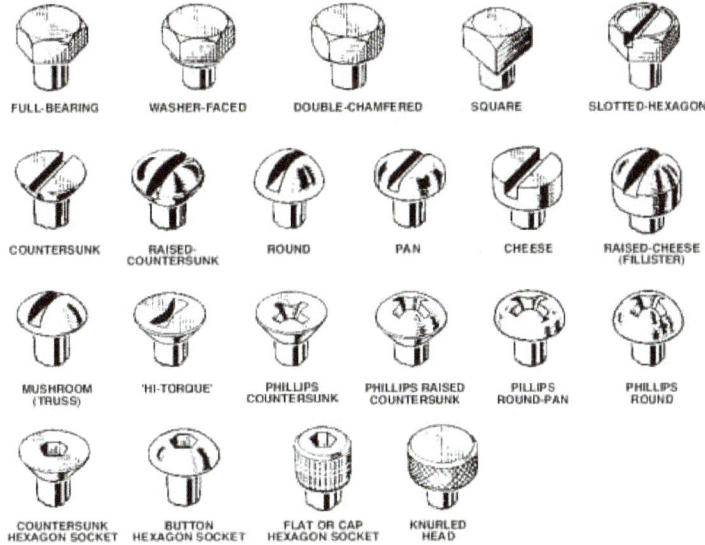

Slotted: The heads of such fasteners are most probably the most ancient and commonly used variety. The linear slot in the screw's head is easily gets along-with any standard screw driver, such as the flat-head screwdriver.

Phillips: These are the enhanced forms of slotted ones. Their cross-shaped laps, which are not continuous up to the edge, will certainly require a Phillips-head screwdriver. The heads of such rivets, featuring a circular shape, enables a larger mating surface, thereby reducing the chances of wear and tear. Plus, they also prevent the chances of slipping. A cross head usually features 2 full-length slots, which can only work with a flat-head screwdriver.

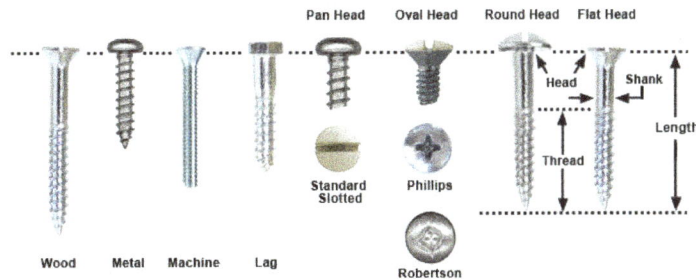

Square: This type is also known as Robertson screw head and features a square dent to reduce the chances of slipping. It requires a special kind of driver for its usage.

Hex: This is again available in two designs. Such types of fasteners either feature a hexagonal recession in the head or do not have grooves at all. The hex socket screw/Allen screw requires an Allen wrench, featuring a hexagonal shaft and the other type of hex screw's head is completely hexagonal in shape. A socket wrench is required to tighten or loosen it.

Apart from being available in above mentioned types, the screws are also available in various materials. While some are made up of steel and stainless steel, there are fasteners, which are made up of brass, nylon, aluminium and are also available with zinc and black-oxide coating.

Types of Screw

There are many different varieties of screw which are selected based on the particular requirement or the materials involved. Some of the most common types include:

Wood Screw

Typically designed with a partially-unthreaded shank and used to attach pieces of timber together.

Concrete Screw

Stainless or carbon steel and used for fastening materials to concrete.

Masonry Screw

Often have a blue coating and are inserted to a pilot hole in masonry.

Double-ended (Dowel) Screw

Have two pointed ends and no head. Often used for making hidden joints between two pieces of timber.

Drive Screw

Smooth, round or mushroom-headed with a reduced diameter shank.

Drywall Screw

Often coated with black phosphate and designed with a bugle head. Used to attach drywall to timber or metal studs.

Eye Bolt

A looped head designed to be used as an attachment point. Also used for attaching wires across building surfaces.

Decking Screw

Longer screws which are used for fastening down deck boards.

Lag screw/Bolt

A heavy-duty fastener.

Chipboard Screw

Often wax-coated and used for fastening down chipboard flooring.

Mirror Screw

Designed with a decorative dome or other cover to conceal the head.

Twinfast Screw

Designed with two threads which enable it to driven twice as fast.

Security Head Screw

Designed with a head that is impossible to reverse, making it suitable for security applications.

Screw Heads

Different types of heads include:

- Pan Head: Rounded, high outer edge with a large surface area.
- Button/dome head: Cylindrical head with a rounded, dome-like top.
- Round head: Dome-shaped and used mainly for decorative purposes.
- Mushroom head: The dome has a lower profile that is designed to prevent tampering.
- Countersunk/flat head: Conical head with a flat outer face and a tapered inner face.
- Oval/raised head: Countersunk bottom and rounded top, often used decoratively.
- Bugle head: A smooth transition from the shank to the angle of the head.
- Cheese head: A disc with a cylindrical outer edge.
- Fillister head: Cylindrical with a slightly convex top surface.
- Flanged head: Can be any style but has the addition, at the base of the head, of an integrated flange which means it does not require a washer.

References

- Fasteners-and-their-types: ispatguru.com, Retrieved 06 May 2018
- Different-types-of-screws-and-their-uses: mutualscrew.com, Retrieved 17 July 2018
- Fastener-threads, hardware: thomasnet.com, Retrieved 24 March 2018
- Threaded-fasteners, products-specs: fastboltcorp.com, Retrieved 14 June 2018
- 4-types-mechanical-fasteners-agricultural-industry: aftagriculturaltourism.com.au, Retrieved 10 May 2018

Maintenance Practices Across Varied Industries

Maintenance typically involves adopting a set of preventive and scheduled practices for the efficient operation of equipment. This chapter is structured in such a way that it will provide an understanding of the maintenance practices required across varied industries, such as aircraft, automotive, bridge, firearm and arm engineering industries.

Aircraft Maintenance

Aircraft Maintenance can be defined in a number of ways and the following may help understand the different aspects:

"Those actions required for restoring or maintaining an item in a serviceable condition including servicing, repair, modification, overhaul, inspection and determination of condition".

"Maintenance is the action necessary to sustain or restore the integrity and performance of the airplane".

"Maintenance is the process of ensuring that a system continually performs its intended function at its designed-in level of reliability and safety".

Aircraft Maintenance is that part of the process of aircraft technical activity which is conducted on aircraft whilst it remains in the line maintenance or base maintenance environment. Aircraft Maintenance is intended to keep the aircraft in a state which will or has enabled a certificate of release to service to be issued. A hangar environment may be available but is often not necessary. The reasons for carrying out maintenance are neatly summarized by.

Aircraft safety: airworthiness at its heart:

1. Keep aircraft in service: Availability, which is of key importance to an operator i.e. the aircraft can meet its schedule.

2. Maximise value of asset (airframe, engines and components): of prime importance to the owner or lessor.

Maintenance will consist of a mixture of Preventive and Corrective work, including precautionary work to ensure that there have been no undetected chance failures. There will be inspection to monitor the progress of wear out processes, in addition to:

- Scheduled or Preventive work to anticipate and prevent failures.

- Unscheduled work – Repair maintenance and On-condition maintenance.

In general terms, for preventive work to be worthwhile, two conditions should be met:

1. The item must be restored to its original reliability after maintenance action, and

2. The cost of maintenance action must be less than the failure it is intended to prevent.

Aircraft maintenance can be divided into three categories:

- Defect rectification (which is the most obvious part – the "fixing" of aircraft).

- Line maintenance.

- Base maintenance (or heavy maintenance or hangar maintenance).

Line Maintenance

Most aircraft (although there are exceptions, like many business jets for example) require line maintenance tasks to be performed quite frequently. In many aircraft types, typical line maintenance tasks would include a daily check (performed anywhere from every 24 to every 48 hours) and a weekly check (every 7-8 days). Apart from that, there may be several OOP (out of phase) maintenance tasks which can be considered to be line maintenance and carried out by a line maintenance provider.

The Part 145 regulation defines line maintenance as any maintenance tasks which can be performed outside of a hangar (under open skies) except for situations in which the weather deteriorates to such conditions, that a hangar becomes mandatory. No wonder – a typical example of a legal definition which means precisely nothing. Technically, one could do any maintenance task under the open sky in the Caribbean assuming the weather is nice and warm and there's no wind.

Many airline professionals have a good "feel" for the line / base distinction and know by heart which is which, especially if they have years of practical experience on a given aircraft type. However, if you're new to the business there's still hope J.

First of all, for some aircraft, the scope of line maintenance is specified in the MPD or MS (Maintenance Planning Document or Maintenance Schedule). Those documents may either bluntly tell you that line maintenance is for every check up to and including the 500 HR A-check, for instance. Or for any A-check. On top of that, the same documentation will specify components which are known as LRU – Line Replaceable Unit. This implies that the replacement of such components can be done during line maintenance. Be careful with oversimplification though – an engine is generally considered to be an LRU and as much as you're allowed to replace it during line maintenance (for instance, if you get FOD – Foreign Object Damage – such as a bird ingestion, at an airport where no hangar facility is available) you should probably stick to a base maintenance facility for this job if there is one nearby.

On the other hand, most modern aircraft tend to define their maintenance programs as standalone tasks rather than complete work packages.

Base Maintenance

The definition of base maintenance is simple – it is all maintenance which does not fall under the line maintenance category. In practice, this will be mainly heavy checks such as C and D checks. During those checks major and minor aircraft systems are being evaluated together with complex and time consuming tasks such as corrosion prevention, structural work, replacement of major components, interior refurbishment, etc. Of course, all this work needs to be done in a hangar and requires quite a bit of planning and a significant dose of cooperation between the airline and the MRO.

The arrangement of base maintenance is not as simple as with line maintenance. Operators are required (by law) to have base maintenance contracts in place for all aircraft they operate. Such contracts need to fulfill the Part 145 contracting / subcontracting requirements and need to be approved by the respective aviation authority.

The way this is generally done is that major MRO's have so-called GTA's – General Terms Agreements – which can be signed by an operator. A GTA does not warrant anything really. It does not ensure that there will be an available slot for your aircraft when you need one, neither does it guarantee any kind of technical support. However, it does present a general framework of the cooperation between the airline and the MRO once a heavy aircraft check is being ordered.

A Quick Method for The Evaluation of Defects:

In most cases, your line maintenance provider will be the first to get the word of the defect (most often through an appropriate entry in the TLB – Technical Log Book – made by the crew or through the results of an inspection like a daily check).

There is no problem in them initially evaluating the defect. Keep in mind that the appropriate Part 145 approvals are needed for performing work on the aircraft (i.e. removing a defect) and not for finding and describing one.

For the sake of this exercise, let's assume that a line maintenance mechanic discovered a fuselage dent on your aircraft. The dent has not yet been described in the dent and buckle chart, which implies that it must have just recently occurred.

Your engineering team (perhaps with the assistance of the line maintenance MRO) should check your SRM – Structure Repair Manual – for allowable defects in the area in question. You checked and there are certain limits within which the defect is acceptable and can be easily released to service (after an appropriate entry in the dent and buckle chart has been made), other limits allow the defect to be released but only for a certain amount of time (say 500 flight hours) by which a permanent repair needs to be done, and everything outside those limits is a "no go" item and requires immediate rectification before next flight.

Here is your first decision – the size of the dent needs to be accurately measured. The measurements you take will define whether your aircraft is airworthy or not, so they need to be really accurate. Mistakes can either endanger flight safety or empty your wallet. In either case, you need to be sure. Can you order your line maintenance provider to do the measurements or should you call a structure expert from a base maintenance facility to do so? There is no definite answer. Ask your line maintenance provider if they feel comfortable with the task – if they have decency, they will give you an honest answer. Furthermore, if you've worked with them for a while you will have your own idea on their abilities and skills. It's really a judgment call – if you're in doubt, have an expert fly in.

The measurements have been made. Bad luck – the dent needs to be fixed immediately. Your line maintenance provider claims to have a good structure engineer on site who can do some of the sheet metal work for you provided you give him instructions from the SRM or obtained from the aircraft manufacturer if necessary. Can you do that? This one is easy – no, you should not allow that. If they are only approved for line maintenance and minor repairs (as is generally the case) they should not be permitted to do the work. You will need to call a base maintenance organization and ask for support. If you're in luck – there will be one at the airport at which your aircraft has been grounded. If not – they need to fly in and do it on site away from base (they most often have their own procedures which allow them to do that).

Aircraft Maintenance Checks

Aircraft maintenance checks are periodic detailed inspections done on all commercial and civil aircraft after a certain period of usage. There are 4 major types of maintenance checks, which are A-Check, B-Check, C-Check, and D-Check.

A-Check is a general maintenance check that is done approximately every 400-600 flight hours. The A-Check needs at least 10 hours to complete an entire process. Services done during the A-Check are, for example, replacement of tires and brakes, fueling, oiling, etc.

B-Check is more sophisticated and time consuming than the A-Check. Performed approximately every 6-8 months, the entire process generally takes around 1-3 days at an airport hangar.

C-Check is an inspection of a large majority of the aircraft's components that is done every 20-24 months. During the C-Check, the aircraft must not leave the maintenance site and remain out of service until the maintenance process is completed. The time required for an entire C-Check process is 1-2 weeks.

D-Check needs at least 2 months to complete. It is the most comprehensive and demanding aircraft maintenance check that happens approximately every 6 years. Also known as "heavy maintenance visit" or HMV, the D-Check usually takes the entire airplane apart for inspection and overhaul.

Automotive Maintenance

The technical category of the maintenance task is comprised of maintenance services and its quality, the methods, resources, materials and control strategies required for maintenance. Following were the observation when the literature was analyzed on this aspect.

The Technical Part

- The maintenance "products". Specification of the different types of services and "products" from the maintenance function. Specification in relation to each plant system.

- Quality of the maintenance "products". Specification of quality of the maintenance jobs. Quality reports, certification documents, decision about maintenance standards, etc.

- Maintenance working methods. Specification of working methods, time standards, relation between maintenance jobs, etc.

- Maintenance resources. Equipment for maintenance, buying maintenance services, information about new equipment, capacity of equipment, usage control, etc.

- Maintenance materials. Inventory planning (spare parts, etc.), warehousing, relation to vendors, etc.

- Controlling maintenance activities. Scheduling of maintenance jobs, progress in work, manpower planning, etc.

The Human Part

- Internal relations in maintenance function. Relation to other departments, corporation and coordination especially to production.

- External relation for the maintenance function. Relation to external parties, especially related to environment and safety. Contact to local authorities, press, labor organization, customer, vendors, neighbors, etc.

- Organization of the maintenance function. Design of the organization, selection of the people, relation between groups of skills, responsibility, and authority.

The Economic Part

- Structure of maintenance. Work breakdown of maintenance, responsibility for work packages, area structure, and relation to accounting system, specification base (drawings, documentation), etc.

- Maintenance economy. Economic control of maintenance: cost estimates, budgets, cash flow, accounting for the maintenance function. Plant investment and financing.

- Production economy. Production economy versus maintenance economy, cost benefit of maintenance.

A traditional approach to maintenance in automobile industries has always been to practice preventive maintenance. But predictive maintenance techniques such as vibration analysis are more useful and are being applied to equipments with the expectation of improving equipment reliability and availability while lowering maintenance costs. Still today many manufacturing systems are mainly maintained by corrective maintenance combined with a certain amount of scheduled preventive maintenance. So far, little has been done on condition monitoring at the robotic assembly lines. Now a day's Total Productivity Maintenance (TPM) is increasingly being applied to improve maintenance function. TPM is a contemporary example of various improvement concepts. Although the concept first appeared more than 30 years ago in Japan, it has only recently found its way to the wider population of companies across Europe, America and Asia, especially within the auto assembly and process industry areas. Further the improvement of maintenance by TPM – can be supported by mobile business in automotive companies. Overall Equipment Effectiveness (OEE), as a measure of TPM implementation effectiveness provides an excellent perspective on production improvement but should be balanced by other, more traditional operational measures, there by retaining an overall perspective of the manufacturing environment.

The industry can also be an important factor in equipment maintenance since the type of equipment, customer demands and strategic uncertainty can differ significantly from industry to industry. In the automobile industry, the large automakers have faced much competition in recent years and thus companies such as Toyota, Ford and Saturn are known to have developed programs such as TPM and its adoption.

For the effective preventive maintenance, maintenance management information system is the foundation for the data that supports the PM's ability to make effective management decisions. It is important to understand that not all MMISs are the same. Thus there are different approaches to maintenance management, but despite advances in computer technology and manufacturing techniques, benchmarking studies of actual maintenance performance signal the need for new, improved methods for analyzing and designing maintenance systems. The application of analytic hierarchy process (AHP) for decision making and to improve the implementation of total productive maintenance is getting momentum in auto industries.

In present scenario TQM and TQC have become the back bone of industries. It would be imprudent even to think of doing only TQC and skipping TPM, or vice versa. A comparative analysis of the total quality control and total productive maintenance paradigms is available which shows that they can be complementary to each other. Also guidelines for nurturing well-balanced organizations based on potentials of both TQC and TPM in the form of Strategic Staircases model has been proposed. The implementation of JIT concept in industrial environment is growing day by day. But JIT requires high machine availability while TQC requires that machines are in excellent working condition, which both in turn requires excellent preventive maintenance.

To avoid failure and improve maintenance mathematical models and robust computer aided simulation have been developed to visualize, analyze and optimize complex maintenance problems in an automotive-manufacturing environment. In the modeling a repairable production unit subject to random failures, which supplies input to a subsequent assembly line, operating according to a just-in-time configuration has been considered for. Setchi & White explained the development environment and stages of creating a Hypermedia Maintenance Manual (HMM) for a manufacturer of automotive equipment. The right policy to counter any mode of failure is that which improves the life cycle profit by reducing the cause of breakdowns in the form of identifying and analyzing different criteria.

The Human Part

The Human category of the maintenance task is comprised of internal relations between the maintenance organization and other departments such as production, external relations to local regulatory authorities, labor organization, vendors and so on, and the design of maintenance organization itself. Human Factor in maintenance management plays an important part for improving efficiency and effectiveness. Productivity improvement by the motor vehicle producers was attained through more efficient utilization of labor ahead of maintenance and other factors. In UK and Japan, shop floor employees' of auto industries work in groups within manufacturing cells; improvement teams formed on the basis of production areas and they meet regularly to discuss problems and their proposed solutions. This team culture together with a flat organizational structure has enabled the companies to develop a good information and communication network. Further there are four management systems which are important for assurance of assembly quality namely production, quality, maintenance and human resources. It is the human factor which affects the quality of assembly most in a discrete part manufacturing assembly lines of auto industry.

Currently the R&D intensity is turning out to be significant in case of automobile industry. Furthermore, there is a significant impact of the R&D intensity and size of the firm on their productivity. Other factors like energy, labor and maintenance also have an impact on the firm's productivity. Thus the evaluation of the maintenance function has identified both weaknesses which have the potential for improvements, and strengths of current maintenance routines. Based on the performance assessment of maintenance department, it is first of all recommended that a brief pre-PM action plan for operators is worked out and implemented to assure its success.

Another concept which is emerging and important in auto industries is the involvement of labor factor in Lean concept. The lean itself is not a single point invention, but the outcome of a dynamic learning process and adapted practices e.g. TPM, JIT etc.. The excellence in equipment maintenance is essential to lean assembly. The Toyota Production System was trapped into the sterile opposition between the empowerment and the management by stress approaches, and has failed to provide a clear understanding of the social and organizational conditions that make this system viable.

The Economic Part

The economic category of maintenance task is comprised of the cost structure of maintenance, the economic control of maintenance (Budgets, cash flow etc.) and production economy. Cost effectiveness of maintenance function is thus an important field of management. Cutting edge auto companies are using many new tools to reduce cost effectively. Since reliability can be designed into the equipment by engineering, demonstrated by operations in careful use of the equipment, and it can only be sustained by maintenance. Businesses cannot afford too little reliability because of high failure costs or too much reliability because of high capital costs. Political and other environment rendered the UK an attractive location for Japanese "transplants" cost effectively. Maintenance management is constantly seeking ways to reduce operating costs, optimize the life cycle cost of components, minimize wasted motion and material and increase availability. Hence to respond to global challenges, the European car companies had to reduce costs, shed labor, rationalize plants, raise productivity and improve their relationships with suppliers in attempts to boost efficiency. Proper maintenance helps to keep the life cycle cost down and ensures proper operations and smooth internal logistics. More and more automobile companies are looking for a customized maintenance.

Maintenance concept is a set of various maintenance interventions (corrective, preventive, condition-based, etc.) and the general structure in which these interventions are brought together. Developing an appropriate maintenance concept is important because of the high direct and indirect costs and because of the operational impact maintenance.

Another concept introduced for reducing cost by many auto industries is agile manufacturing system integrated with maintenance concept. Agile manufacturing systems from an automotive industry maintenance perspective seem to meet the promise of rapid and cost-effective response to manufacturing. A Latest development in cost reduction effort is the near net shape technology. Cominotti & Gentili reported that Near Net Shape technology in automotive industry helps in achieving cost reduction, reduction of process variability, quality improvement in the finished product and simplifies maintenance. The partial maintenance productivity goal for a firm is to maximize its maintenance productivity in economic terms and should aim at producing any level of output which is decided upon at minimum maintenance cost with respect to the production systems state. These imputed maintenance costs do not have to be calculated separately, but emerge as a by-product of finding a high productivity index.

Bridge Maintenance

Bridge Repair can be divided into three categories: Maintenance, Rehabilitation, and Replacement.

Maintenance Activities Include

- Scheduled inspections of all moveable and stationary bridges and culverts.

- Repair and Replacement of bridge railing, decks, approaches and substructures.

- Removal of Drift and Debris.

- Erosion protection.

- Moveable Bridge Operation and Maintenance.

- Guardrail and Retaining Wall Repair and Maintenance.

Rehabilitation activities include stringer replacement, deck replacement, bridge painting, etc. Several annual contracts are described below:

- Bridge Repair Project

The County typically has one bridge repair project annually. These types of projects are typically under $100,000.

Bridge Painting Project

The County typically has one bridge painting project annually.

Reconstruction is the complete reconstruction of the bridge structural section or replacement with a large capacity culvert. Examples include:

- Bridge Replacement Project

The bridge replacement project involves removing the existing bridge and replacing it with a new concrete bridge or culvert, re-grading the existing channel, providing slope protection, replacing the existing road surface to match the existing surface, etc.

- Culvert Replacement Project

At this time the County replaces culverts in-house or using a minor contract.

Techniques

Infrared Thermography and Ground-penetrating Radar

Infrared thermography and ground-penetrating radar have been developed to locate voids and delaminations in concrete structures such as bridge decks, highways and airport pavements. Being able to locate voids and delaminations means the structural maintenance engineer can measure the actual cracking and weakening of concrete pavements before catastrophic failures can occur.

Concrete objects, such as bridges, emit energy based upon the absolute temperature of its surfaces and the surface temperatures are dependent upon the internal conditions of the concrete. These internal conditions can include physical conditions like:

1. Density changes in concrete

2. Voids caused by erosion beneath the concrete slabs

3. Horizontal delaminations caused by rust expansion of rusting internal reinforcing steel.

Infrared thermographic radiometer or "IR Imager" locates these anomalous thermal conditions. This device can measure hundreds of thousands of individual temperature points per second and convert this data to thermal maps or temperature images of the concrete. By locating anomalous areas, or temperature patterns which differ from the background "norm" on these images, trained engineers can locate the exact anomalous areas that could lead to catastrophic failure of concrete and its supporting soil and backfill systems.

Ground-penetrating radar gives information valuable in determining such characteristics as: target material, voids, fluids, soil or backfill strata, and quantity of reinforcing steel present.

Magnetometer

Magnetometers are instruments designed to locate ferrous materials. It can detect iron containing materials to a maximum depth of approximately 10 feet. This is useful for locating dowel pins or determining if reinforcing steel exists.

Pachometer

This device is designed to specifically locate reinforcing steel in concrete and to assist in the determination of the size of the hidden reinforcing steel.

Firearm Maintenance

Everyone has an opinion about how to maintain firearms, including what to use to clean them from bore solvents and miracle oils to patches, brushes and ultrasonic machines. Unfortunately, not enough officers take the necessary care to ensure optimal performance from their weapons, nor do they realize the potential hazards of a poorly maintained firearm.

The Basic Cleaning Kit

Cleaning Solvents

It all starts with a good can of cleaning solvent. Cleaning solvents will help remove residue from your barrel, and they are best used in conjunction with cotton cleaning patches. Note that cleaning solvent and gun oils are not the same thing. You start with cleaning solvents and finish with gun oil. Also remember that moderation is key. You should only use a minimum amount of cleaning solvent as excess solvent can negatively impact key firearm parts like the trigger. There are also toxicity concerns with some cleaning solvents. While the harsher, more toxic varieties can clean better, they can also harm your skin and lungs if used improperly. Biodegradable alternatives are worth a try to protect the environment and safeguard your health.

Gun Oil

Gun oil is the other necessary can of liquid that you need to clean your firearm. Gun oil helps lubricate your firearm and protect it from the onset of rust. You'll find all in one oils and solvents and lubrication specific oils. Which one to use is a matter of preference, but the most comprehensive solution is usually a combination of a cleaning solvent and a separate gun oil.

Cleaning Rod

A good cleaning rod is an essential part of any cleaning kit. They are usually separated into a few segments and they screw together to give you the appropriate length for your chosen firearm. They come in several different materials like aluminum, fiber glass, carbon fiber, and brass. Brass usually represents the strongest option although they will be heavier.

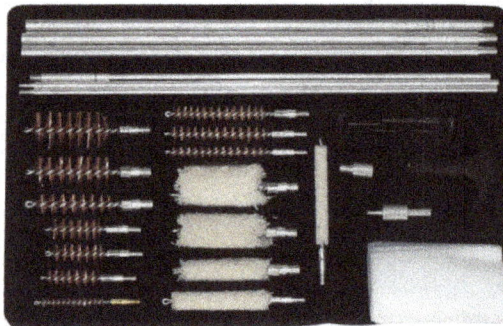

Cleaning kit consisting of cleaning rod, brushes, and cleaning mops.

Bore Brush

A good quality bore brush will help remove residue and keep your barrel as clean as possible. Brushes are caliber specific, so for the best performance, use only approved caliber brushes in your firearm barrel. You can also find some universal brushes out there depending on how customized you want your kit to be. Brushes come in a variety of materials but again brass is the most popular material.

Patch Holder

An essential part of the cleaning process involves a patch holder that holds cotton cleaning patches. Patch holders are configured to fit a variety of barrels, but make sure the patch holder you select is approved for your chosen firearm caliber. Patch holders are usually made of brass, and they will attach to your cleaning rod just like a bore brush. An alternative to a patch holder is a cleaning jag that acts more like a pushing device for the cleaning patch. Jags are slightly less wide than the caliber barrel you're cleaning, and you'll need to get one specifically designed for each barrel you plan to clean.

Cotton Patches

Be sure to stock up on a good number of cotton patches. They are great for running solvent through your barrel, but since they gather residue, you can only use them once before discarding them.

Luster Cloth

A luster cloth is a specialized cloth treated with a silicon lubricant. It is the perfect material for use as a wiped down cloth for the outside of the firearm. A little dab of gun oil is used with a luster cloth in the final part of the cleaning process.

Extras / Advanced Cleaning Materials

Cotton / Wool Mops

Realistically, you can get away with just using cotton patches, but a good cotton mop provides the best option for applying gun oil to the barrel once cleaning is completed. Cotton mops can be sized for a specific caliber barrel, and they will only apply a light coat of oil to the barrel.

Muzzle Guards

The most seasoned firearm owners always caution owners about cleaning rod damage. For only a few extra dollars, you can add a muzzle guard and protect the muzzle crown from unnecessary damage. Muzzle guards also keep the cleaning rod centered, and they are usually composed of brass materials.

Firearm Cleaning Mat

If you're cleaning a firearm with many advanced parts to disassemble and reassemble, then a firearm mat can be a great investment. A quality firearm mat serves several purposes. It provides an ideal cleaning / work surface, it is non-slip so it keeps gun parts in place, and it is printed with a diagram of your chosen weapon so you have a guide to where every part fits. Firearm mats are also resistant to cleaning chemicals, and even if you don't necessarily need the diagram, they are a good investment for firearm cleaning around the house.

Safety Concerns and Materials

Let's start by stating the important, but obvious. When you clean your firearm, you should do it in a well-lit and ventilated area, away from open flames. You should also remove any ammunition that may be present. Make sure the firearm is unloaded before disassembly. You're going to want some old T-shirts, some q-tips, that old toothbrush that you needed to replace (you get a new one every dental visit anyway) and your cleaning kit consisting of patches, bore/chamber brush, punch rods cleaning solvent and lubricant. You may choose to use one of those nifty bore snakes, which I find work well. After all that, put on some of your favorite music and start your relaxation therapy.

A field strip of a 1911 in preparation for cleaning

If you are unsure, make sure to reference your owner's manual when you disassemble your firearm. If you don't have an owner's manual, there are some great tutorials on YouTube these days. There are also some horrible tutorials on YouTube, so make sure to evaluate the source of your information. Another option is to visit the manufacturer's webpage itself. Often, you can download the owner's manual directly from the website. Not only does the owner's manual contain assembly and disassembly instructions, it often provides information on which points to apply oil.

Please remember that when applying oil to your firearm, doesn't overdo it. If one drop of oil is sufficient, 50 drops are not 50 times better. Applying oil to areas not requiring oil, or applying too much oil is detrimental to the function of your firearm. Oil attracts debris such as sand, dust, dirt lint, powder, etc. This can create a dirty, sludge-like build up that could cause a malfunction. Guns run well with light oil in key spots designated in your owner's manual. Without this instruction, key points such as the rails in a semi-automatic and a light coat on the outside of the barrel are all that is really needed. In absence of specific points named by the manufacturer (usually involving removal of the grips and lightly oiling the internals), most revolvers only need to be cleaned well and wiped down with a lightly oiled cloth.

Please remember that when applying oil to your firearm, doesn't overdo it. If one drop of oil is sufficient, 50 drops are not 50 times better. Applying oil to areas not requiring oil or applying too much oil is detrimental to the function of your firearm. Oil attracts debris such as sand, dust, dirt lint, powder, etc. This can create a dirty, sludge-like build up that could cause a malfunction. Guns run well with light oil in key spots designated in your owner's manual. Without this instruction, key points such as the rails in a semi-automatic and a light coat on the outside of the barrel are all that is really needed. In absence of specific points named by the manufacturer (usually involving removal of the grips and lightly oiling the internals), most revolvers only need to be cleaned well and wiped down with a lightly oiled cloth.

There are many cleaning products on the market that promise outstanding results. Ask around and find what people seem to like, and it may work for you. There are some cleaning products that are not solvent based. These are a good option if you do not like the caustic properties or strong odor of solvents. These also have the benefit of not drying out your hands or potentially damaging plastic parts or accessories. Whatever product you choose, there are key points you want to clean. In a semi-automatic.

In a semi-automatic pistol, these are the feed ramp, chamber, slide rails, trigger group area, and magazine well. For most areas, scrubbing is not necessary. Carbon and buildup tend to form on the feed ramp and chamber area, so if it's been a while since these areas were cleaned, it may need a good scrubbing. If your toothbrush isn't enough to remove some stubborn carbon build up in these areas, a brass wire brush can be used. Make sure to clean your magazine well. This area needs to be clear of any debris to allow the magazine to slide in and out properly. While you are at it, ensure you clean your magazines also. Especially if you have been running reloading drills and your mags have spent some time being dropped on the ground. Remove the base plate and clean the inside. Clean the follower and put it all back together. Do not put oil inside your magazines. This will surely attract debris and adversely effect the function of your magazines. For revolvers, the cleaning task is easy. The chamber,

Make sure to clean your magazine well. This area needs to be clear of any debris to allow the magazine to slide in and out properly. While you are at it, ensure you clean your magazines also. Especially if you have been running reloading drills and your mags have spent some time being dropped on the ground. Remove the base plate and clean the inside. Clean the follower and put it all back together. Do not put oil inside your magazines. This will surely attract debris and adversely affect the function of your magazines.

For revolvers, the cleaning task is easy. The chamber, cylinder, and barrel all need a good cleaning. When cleaning the barrel of any firearm, it is preferable to clean in the same direction the bullet travels. This is mainly because the area at the very end of the barrel, the crown or muzzle, could be damaged if you are using a metal cleaning rod inserted from the muzzle end. If you are using plastic cleaning rods or using a bore snake it really is not going to matter which direction you clean your barrel. In some cases like with some snub-nosed revolvers, it is nearly impossibly to clean it without going from the muzzle side.

Cleaning your barrel is simple and satisfying. First clean with your cleaning agent, and a barrel brush made for your caliber. After passing the brush several times, pass a cleaning rod with a patch attached through your barrel. A patch is simply a small, clean, cotton cloth. Pass the patches through your barrel until they come out clean. Look down your barrel in the light and observe the lands and groves happily shining away.

Final Steps and Inspection

After cleaning your firearm, it is important to put it back the proper way and function check it. Cycle your slide (semiautomatic) and pull your trigger. Ensure the firearm is functioning i.e. hammer falling, striker releasing and resetting, etc. If you have dummy rounds/snap caps you can load them into your firearm and ensure the firearm is feeding, extracting and ejecting correctly. Lastly, put a few drops of oil in a cloth and wipe the firearm down. Do not put too much oil on your firearm or it will be slippery and difficult to handle. You don't want to have to handle an oily, slippery weapon, or have it in your expensive, new leather holster.

The Importance of Cleaning your Firearm

Keeping your firearms clean is important for multiple reasons. Primarily, a clean handgun is a safe handgun. When firing, powder residue and grimy substances are left on the exterior, in the action and in and on the barrel. Over time, this buildup can prevent your firearm from operating the way it is intended to. The results can be both dangerous and cause unreliable, unpredictable performance.

Another reason to keep your handgun clean is to improve its useful life. A handgun that is properly treated, cleaned and cared for can last for decades, while, a firearm that is rarely cleaned or maintained could rapidly deteriorate.

How to Clean your Firearm

Cleaning your firearm can be easy and enjoyable if you follow the simple steps laid out here:

- Buy a cleaning kit: To get started, you will need some very specific cleaning supplies and tools. Most sporting goods stores sell preassembled cleaning kits, but check to ensure everything you need is included. This includes cleaning solvent, lubricant, cleaning rods, bore brushes, cotton swabs, polishing cloths and patches.

- Unload the firearm: You should never clean your firearm while it is loaded. Start by removing the magazine and double-check to ensure there are no rounds in the chamber. A good rule of thumb is to have no ammunition present when and where you clean.

- Disassemble the firearm: Your handgun will come with specific manufacturer instructions on how to disassemble it. Carefully follow these instructions and remove the various parts for cleaning.

- Clean the barrel: One of the most important parts to clean is the barrel. To do this, use the brushes and patches included in your cleaning kit to remove grime and lubricate.

- Clean all parts of the action: Be sure to remove any buildup of dirt or residue down to the finished metal or polymer surface.

- Lubricate the action: Next, apply some solvent to the brush and lubricate the moving parts of the action. A small amount goes a long way, so only use a little for best results.

- Reassemble the firearm: After cleaning the barrel and action, reassemble the firearm according to manufacturer instructions.

- Wipe down the outside. The final step is to clean the outside of the firearm with an appropriate cloth to remove oils, dirt and other foreign substances.

Army Engineering Maintenance

Army planners agree that a transformed materiel maintenance system must substantially improve equipment reliability, reduce the size of logistics support elements, and enhance maintenance responsiveness. However, despite various programs and initiatives committed specifically to developing new maintenance concepts, processes, and technologies, the best way to proceed with achieving these goals has not been defined clearly. Attempts to make equipment sustainment equal in importance to other war-fighting considerations have not yet produced meaningful results. In fact, Army maintenance procedures have changed very little over the past decade or so, and our Soldiers are still encumbered with overly bureaucratic maintenance policies, archaic logistics information systems, and equipment that may have been designed and engineered more for "manufacturability" than maintainability. Simply stated, the Army cannot expect to transform itself successfully without a genuine, conspicuous, and quantifiable commitment to reinvent maintenance.

1. Accelerate introduction of embedded diagnostics and prognostics: Embedded diagnostic and embedded prognostic (ED/EP) systems truly are the technological "heart" of a transformed maintenance system. Much more than an on-board troubleshooting tool, the ED/EP system also must be the primary conduit for many other sustainment functions, such as joint logistics information control, digital preventive maintenance checks and services, automated status reporting, platform-based parts requisitioning, remote diagnostics, telemaintenance, vehicle configuration management, component life-history recording, and embedded just-in-time maintenance training. Several ex- periments have successfully demonstrated the value of this multifunctional approach to ED/EP, including the Army G–4's visionary Common Logistics Operating Environment (CLOE) initiative, which is now the standard ED/EP operational architecture for connecting logisticians. [CLOE guides the Army's vision for developing a technology-enabled force equipped with self-diagnosing platforms that interact with a networked sustainment infrastructure.

Maintenance transformation does not depend solely on innovative technology. Instead, real transformation results from profound cultural change that is enabled by technology. Therefore, the greatest return on investment from accelerated fielding of multifunctional ED/EP systems (for both Current Force and Future Force platforms) will be the creation of an enormous window of opportunity for modernizing logistics policies and procedures.

2. Update maintenance processes using CBM+ as the central theme: Because assured mobility is so crucial to the operational effectiveness of our Future Force, we must give leaders the option of replacing components before the actual point of failure. The Army must develop a transformed logistics system that blends conventional maintenance techniques with Department of Defense Condition-Based Maintenance Plus (CBM+) guidelines. Moving from a fault-based maintenance philosophy to one that is anticipatory, proactive, and reliability centered will decrease the battlefield maintenance workload, boost reliability during combat pulses, and reduce costs by avoiding catastrophic failures.

3. Adopt a NASA mentality for future ground platforms: Future ground platforms must be designed and engineered for improved maintainability, rapid repair, nominal tool requirements, redundancy, system bypass capability, and maximum use of plug-and-play modular components. This methodology, often called "pit stop engineering," also can be compared to the design philosophies of the National Aeronautics and Space Administration (NASA) manned space program. NASA designs spacecraft using strict reliability standards and incorporating multiple, redundant systems for continued operation, even during failures. If the Army wants to conduct sustained battlefield operations with minimal logistics support, it must invest in combat platforms that include at least some measure of engineering borrowed from the space program.

4. Increase maintenance performed by equipment operators and crews: The noncontiguous battlefield anticipated for future conflicts restricts the ability of logisticians to project maintenance support. With combat repair teams operating independently over extended distances, vehicle crews experiencing maintenance problems cannot always expect a timely response from field maintenance personnel. In some cases, a crew's survival may depend on its ability to diagnose faults and make repairs quickly.

With this in mind, the Army's combat maintainer model was introduced as a central feature of the Army Training and Doctrine Command-approved Stryker advanced maintenance concept. Patterned after the combat lifesaver model of field medical support, the combat maintainer program expands maintenance effectiveness and combat self-sufficiency by training vehicle crewmembers to perform selected mission-critical equipment repair tasks, basic troubleshooting, self- or like-vehicle recovery, and limited battlefield damage assessment and repair procedures.

5. Establish sense-and-respond processes for repair parts supply: The Office of Force Transformation's sense-and-respond logistics project holds great potential as a principal enabler for rapid distribution of mission-critical repair parts. The two primary repair parts management challenges in today's multidimensional combat environment are inaccurate anticipation of demands and sluggish battlefield distribution. Multifunctional ED/EP systems and sense-and-respond logistics can help mitigate these challenges through dynamic networking of dispersed logistics resources. Fundamentally, sense-and-respond logistics considers all repair parts, regardless of where they are stored or to which unit they belong, as a common pool that can be requested by any network user and delivered by any available asset. Under this concept, support roles are flexible and

continuously adaptive, and logistics customers may be tasked periodically to function as logistics providers. Ultimately, sense-and-respond logistics processes will enhance the availability of repair parts across the battlespace without requiring a corresponding increase in logistics support structure.

6. Eliminate the notion of "levels of maintenance": In the purest terms, maintenance can be viewed strictly as another sustainment function that the Army must perform, regardless of "who, what, when, why, where, or how." All maintenance tasks could be consolidated into a single category, and it would no longer be necessary to describe the Army maintenance system using obsolete terms from the linear battlefield, such as "levels." While the Army's transition from four to two levels of maintenance has produced some benefits, the traditional practice of pigeonholing tasks into rigid columns on a maintenance allocation chart eventually can be replaced with a unified and highly adaptable maintenance philosophy that eliminates levels altogether.

7. Allocate maintenance tasks using decision logic. Once echeloning of the Army's maintenance system is abolished, responsibility for performing maintenance tasks can be determined by using a decision chart, with training and resources as the main considerations for task accomplishment. All Soldiers can be trained to apply task decision logic and quickly evaluate maintenance factors on the decision chart before proceeding with equipment repairs. Thus, if all of the decision chart requirements are met, the task is performed; if any of the requirements are not fulfilled, the task must be deferred or reassigned to another maintenance element.

Maintenance Task Decision Chart							
	TRAINING	TIME	PERSONNEL	TOOLS & TMDE	REPAIR PARTS	EXPENDABLES	IETMS	
YES	☐	☐	☐	☐	☐	☐	☐	PERFORM TASK
NO	☐	☐	☐	☐	☐	☐	☐	DEFER TASK OR EVACUATE

Evaluate each of the above maintenance factors before making maintenance task decisions.

An example of a maintenance task decision chart.

8. Develop a single, all-encompassing interactive electronic technical manual for each future platform: Eventually, on-board, interactive electronic technical manuals can be consolidated into a single reference tool (one manual for each platform or equipment item), and separate manuals for different levels of maintenance will be unnecessary. Future interactive electronic technical manuals also must include a master task list, similar to that found in commercial automotive service manuals, with detailed information that corresponds to the seven task-evaluation factors on the maintenance task decision chart. (See the example above.)

9. Purge the term "mechanic" from the Army's vocabulary: Perhaps the introductory paragraph from the Web page of the automotive technology program at South Puget Sound Community

College in Olympia, Washington, best describes the changing nature of automotive service and repair: "A mechanic goes after your car with a hammer. An automotive technician talks to your car with a computer." Since digitally controlled systems are so commonplace in modern automotive designs, the knowledge, skills, and abilities of today's automotive service technician are distinctly different from those of yesterday's "grease monkey." Similarly, modern Army equipment has increased in sophistication to the point that the term "mechanic" does not accurately reflect the depth of technical expertise required to maintain our newest ground platforms. Because our professional maintenance Soldiers' roles on the future battlefield will be even more critical than they are now, calling them "technicians" is an important first step in changing the way we recruit, train, deploy, and retain them.

10. Revamp and certify maintenance training programs: When maintenance levels are eliminated, task allocation is linked to resources, and mechanics are replaced with technicians, automotive maintenance training can be transformed into three exportable modules: an entry course for equipment operators and crews, a basic course for new Ordnance Corps mechanical maintenance enlistees, and an advanced course for senior technicians. Eventually, all Soldiers will take the entry course, regardless of their primary military occupational specialties, to support distributive maintenance concepts by increasing operator and crew maintenance responsibilities.

References

- Aircraft-Maintenance: skybrary.aero, Retrieved 10 March 2018

- Aircraft-maintenance-line-base-and-defects: airlinebasics.com, Retrieved 24 May 2018

- Types-of-aircraft-maintenance-checks: taithailand.com, Retrieved 25 April 2018

- Bridge-Repair: sacdot.com, Retrieved 12 May 2018

- The-importance-of-firearms-maintenance: concealedcarry.com, Retrieved 20 June 2018

- Handgun-maintenance-7-steps-to-easy-cleaning: springfield-armory.com, Retrieved 10 July 2018

Permissions

Index

A

Abrasive Wear, 50, 87-88
Adhesive Wear, 49-50, 88
Air Bearing, 100, 122, 125-126
Aircraft Maintenance, 164-165, 167
Angular Contact, 110-112
Army Engineering Maintenance, 179
Automotive Maintenance, 167, 182
Axial Load, 110-112, 117, 121

B

Ball Bearing, 100, 103, 109-114, 116, 118, 120
Base Maintenance, 164-167
Bearing Failure, 38, 42-46
Bending Fatigue, 57-58, 122
Bore Brush, 174
Bridge Maintenance, 171
Bridge Painting, 171
Brinelling, 43, 45, 103, 115

C

Cleaning Rod, 174, 176, 178
Cleaning Solvent, 173, 176, 178
Combustible Dust, 59-60, 63
Composite Bearing, 107
Corrective Maintenance, 2-3, 5, 13-14, 168
Criticality Analysis, 37, 63-65

D

Deep-groove Ball Bearing, 109
Destructive Pitting, 52-53
Dry Friction, 75-76, 82

F

Failure Analysis, 17-18, 20, 37-41
Failure Mode, 37, 63-68
Failure Rate, 37, 63, 67-70
Fastener Thread, 154-155, 159
Firearm Maintenance, 173
Fluid Friction, 75, 82, 92
Fmeca, 63-68
Fretting Corrosion, 45, 47, 89

G

Grease Fitting, 73, 97-98, 105
Gun Oil, 173, 175

I

Indentation, 45, 151

K

Kinetic Friction, 73, 75-80, 85

L

Lubricated Friction, 75, 82

M

Machine Element, 25, 31-32, 34-35
Maintenance Engineering, 1, 23, 35
Maintenance Management, 1, 3, 7, 14, 24, 169-170
Maintenance Model, 20-22
Maximum Load, 108, 115-118, 121, 151
Mechanical Fastener, 160
Modern Materials, 115, 148

N

Normal Force, 76-81, 85, 102

P

Plain Bearing, 100, 107-108, 126, 128, 132
Planned Maintenance, 7, 10-14
Planned Maintenance System, 10, 12-13
Plastic Flow, 55-56
Predictive Maintenance, 5, 14-17, 21, 65, 168
Preventive Maintenance, 3, 6-10, 13-14, 16-18, 168-169, 179
Productive Maintenance, 12, 18, 169

R

Reducing Friction, 84, 92-93, 100, 102
Roll Threading, 160
Rolled Thread, 160
Roller Bearing, 100, 115, 119-122
Roller Thrust Bearing, 115
Rolling Element Bearing, 43, 103, 106-107, 125-126
Rolling Friction, 44, 59, 74, 83-84
Running Rigging, 142, 146-148

S

Sailboat Rigging, 146
Screw Head, 161
Sliding Friction, 74, 76, 78, 80, 84
Smearing, 44, 46-47, 106
Spalling, 43-46, 53
Spiral Bevel Pinion, 50, 56
Spiral Groove Bearing, 100, 126, 133
Spiral-groove Journal Bearing, 129-132, 134-135, 137-139
Standing Rigging, 142, 146, 148-149

Static Friction, 73-74, 76-80, 85, 100
Synthetic Rigging, 142-146

T

Threaded Fastener, 154, 158
Thrust Ball Bearing, 109, 111
Tooth Breakage, 56-59

V

Viscous Medium, 126, 128, 130

www.ingramcontent.com/pod-product-compliance
Lightning Source LLC
Chambersburg PA
CBHW082011190326
41458CB00010B/3150